THE LORDS OF LIMIT

By the same author

*

Geoffrey Hill

The Lords of Limit

ESSAYS ON LITERATURE
AND IDEAS

New York
OXFORD UNIVERSITY PRESS
1984

Copyright © 1984 by Geoffrey Hill
First published in Great Britain by André Deutsch Limited, 1984
First published in the United States
by Oxford University Press, New York, 1984
First issued as an Oxford University Press paperback, 1984

Library of Congress Cataloging in Publication Data
Hill, Geoffrey.
The Lords of limit.
1. English poetry—History and criticism—
Addresses, essays, lectures. 2. American poetry—History
and criticism—Addresses, essays, lectures.
3. Poetry—Addresses, essays, lectures. I. Title.
PR503.H54 1984 801'.951 84-14751
ISBN 0-19-503516-X
ISBN 0-19-503517-8 (pbk.)

Printing (last digit) 9 8 7 6 5 4 3 2 1

Printed in the United States of America

To
J. P. Mann

Contents

Acknowledgements

Eight of the nine essays contained in this book have been previously published. I herewith make grateful acknowledgement to the editors and publishers of the respective books and journals and to the Council of the Royal Society of Literature.

1 'Poetry as "Menace" and "Atonement"' was delivered as an inaugural lecture in the University of Leeds on 5 December 1977. It was published in *The University of Leeds Review*, vol. 21 (1978).

2 'The Absolute Reasonableness of Robert Southwell' was delivered as the Joseph Bard Memorial Lecture to the Royal Society of Literature, London, on 17 May 1979. This is its first appearance in print.

3 '"The World's Proportion": Jonson's Dramatic Poetry in *Sejanus* and *Catiline*' first appeared in *Jacobean Theatre*, edited by John Russell Brown and Bernard Harris, Stratford-upon-Avon Studies, 1 (London, Edward Arnold, 1960). Some changes of wording were made in the 1972 reprint.

4 '"The True Conduct of Human Judgment": Some Observations on *Cymbeline*' was contributed to *The Morality of Art: Essays Presented to G. Wilson Knight by his Colleagues and Friends*, edited by D. W. Jefferson (London, Routledge and Kegan Paul, 1969).

5 'Jonathan Swift: the Poetry of "Reaction"' was first published in *The World of Jonathan Swift*, edited by Brian Vickers (Oxford, Basil Blackwell, 1968).

6 'Redeeming the Time' first appeared in *Agenda*, vol. 10 no. 4 – vol. 11 no. 1 (Autumn-Winter 1972–3), edited by William Cookson and Peter Dale.

7 '"Perplexed Persistence": the Exemplary Failure of T. H. Green' was first published in *Poetry Nation* (latterly *PN Review*), no. 4 (1975), edited by C. B. Cox and Michael Schmidt.

8 'What Devil Has Got Into John Ransom?' was delivered as the Judith E. Wilson Lecture on Poetry in the University of Cambridge, on 15 February 1980. It was published in *Grand Street*, (Summer 1983), edited by Ben Sonnenberg.

9 'Our Word Is Our Bond' first appeared in *Agenda*, Vol. 21 no. 1 (Spring 1983), edited by William Cookson and Peter Dale.

O Lords of Limit, training dark and light . . .

W. H. AUDEN

Quintilian calls him [Horace] 'felicissime audax', and Petronius refers to his 'curiosa felicitas' or 'studied felicity'. *Oxford Companion to Classical Literature*

And for this reason we call the doctrine of *the things that can be and go wrong* on the occasion of such utterances, the doctrine of the *Infelicities.* J. L. AUSTIN, *How to Do Things with Words*

It is always a significant question to ask about any philosopher: what is he afraid of? IRIS MURDOCH, *On 'God' and 'Good'*

. . . the Igbo believe that when a man says yes his *chi* will also agree; but not always. Sometimes a man may struggle with all his power and say yes most emphatically and yet nothing he attempts will succeed. Quite simply the Igbo say of such a man: *Chie ekwero*, his *chi* does not agree. Now, this could mean one of two things: either the man has a particularly intransigent *chi* or else it is the man himself attempting too late to alter that primordial bargain he had willingly struck with his *chi*, saying yes now when his first unalterable word had been no, forgetting that 'the first word gets to Chukwu's house'. CHINUA ACHEBE, *Morning Yet on Creation Day*

ONE

Poetry as 'Menace' and 'Atonement'

Thus my noblest capacity becomes my deepest
perplexity; my noblest opportunity, my uttermost
distress; my noblest gift, my darkest menace.

My title may well strike you as exemplary in a fashion, being at once assertive and non-committal. The quotation-marks around 'menace' and 'atonement' look a bit like raised eyebrows. 'Menace' from what, and to whom? 'Atonement' by whom, and for what? Is one perhaps offering to atone for the menace of one's own jargon? In fact, though my title may appear 'challenging', it presents little more than a conflation of two modernist clichés. That it does so is an act of choice but the choice is exercised in order to demonstrate the closeness of a constraint. Behind the façade of challenge is the real challenge: that of resisting the attraction of terminology itself, a power at once supportive and coercive. There is for me also the challenge of the occasion, and a matter of decorum. I cannot disguise from myself the awareness that I have been drawn towards my present theme by way of the technical and metaphysical problems which I have encountered as a practitioner of verse. To what extent I should disguise this awareness from my audience is a question that causes me some perplexity. I am committed to speak, *ex cathedra*, as a professor of English Literature, not as a poet in residence. My distinguished predecessor, A. Norman Jeffares, is a scholar with an unrivalled knowledge of the life and work of one of the greatest of modern poets, W. B. Yeats. In taking poetry, and particularly modern poetry, as my own theme I shrink from any implication of special pleading, disdain the 'confessional mode' as currently practised, but distrust enigmas. That I have had some practice in the making of verse is evidence to be noted, I think; if only as a glint of improper goliardic song in the margin of a proper gospel.

Milton's dictum that poetry, though 'lesse suttle and fine', is 'more simple, sensuous and passionate' than rhetoric is a saying to which I am sympathetically inclined. Ideally my thesis would be equally deserving of

1

sympathy. That it is here presented garnished and groaning with obliquities is due less to a simple sensuous and passionate wilfulness than to an obvious yet crucial fact. Language, the element in which a poet works, is also the medium through which judgments upon his work are made. That common-place image, founded upon the unfinished statues of Michelangelo, 'mighty figures straining to free themselves from the imprisoning marble', has never struck me as being an ideal image for sculpture itself; it seems more to embody the nature and condition of those arts which are composed of words. The arts which use language are the most impure of arts, though I do not deny that those who speak of 'pure poetry' are attempting, however in-adequately, to record the impact of a real effect. The poet will occasionally, in the act of writing a poem, experience a sense of pure fulfilment which might too easily and too subjectively be misconstrued as the attainment of objective perfection. It seems less fanciful to maintain that, however much a poem is shaped and finished, it remains to some extent within the 'imprison-ing marble' of a quotidian shapelessness and imperfection. At the same time I would claim the utmost significance for matters of technique and I take no cynical view of those rare moments in which the inertia of language, which is also the coercive force of language, seems to have been overcome.

Ideally, as I have already implied, my theme would be simple; simply this: that the technical perfecting of a poem is an act of atonement, in the radical etymological sense – an act of at-one-ment, a setting at one, a bringing into concord, a reconciling, a uniting in harmony; and that this act of atonement is described with beautiful finality by two modern poets: by W. B. Yeats when he writes in a letter of September 1936, to Dorothy Wellesley, that 'a poem comes right with a click like a closing box' and by T. S. Eliot in his essay of 1953, 'The Three Voices of Poetry':

> when the words are finally arranged in the right way – or in what he comes to accept as the best arrangement he can find – [the poet] may experience a moment of exhaustion, of appeasement, of absolution, and of something very near annihilation, which is in itself indescribable.

Anyone who has experienced that moment in which a poem 'comes right' must, I believe, give instinctive assent to such statements. And yet, in admitting this word 'instinctive', do I not put my argument in jeopardy and betray my deepest conviction? For it is not my intention to say anything which could either excite or placate those who associate creativity with random spontaneity and who regard form and structure as instruments of repression and constraint. It is as well to be reminded that my phrase was 'instinctive *assent*'; and if 'instinct' is a 'natural or spontaneous tendency or inclination', 'assent' is 'agreement with a statement . . . or proposal that does not concern oneself'. From the depths of the self we rise to a concurrence with that which is not-self. For so I read those words of Pound: 'The poet's job is to *define* and yet again define till the detail of surface is in accord with the root in justice'.

I am attempting to convey, through these preliminary remarks, my belief that a debate of this nature is committed to a form of mimesis. The speaker must submit to an exemplary ordeal, analogous to that ordeal which Empson disarmingly calls 'the effort of writing a good bit of verse'. 'Mimesis', though, is an alluring term and exemplary ordeals are supposed to be ascetic. Define and yet again define. When Auerbach, in his book *Mimesis*, refers to a 'method of posing the problem so that the desired solution is contained in the very way in which the problem is posed' he acknowledges a pattern which is both austere and seductive. In posing the problem we 'show what it is like' to come up against rawness and contingency but not for a moment do we seriously put our mastery in hazard. When D. M. Mac-Kinnon, on the other hand, remarks that Plato may have recognized, in the life and death of Socrates, 'a concretion, one might say a *mimesis*, of the way in which things ultimately are' we are possibly shaken out of our self-containment, our passionate attachment to those forms of hermetic mastery which must be so rebuked by life. But Romantic art is thoroughly familiar with the reproaches of life. Accusation, self-accusation, are the very life-blood of its most assured rhetoric: As Yeats puts it, in his poem 'The Circus Animals' Desertion':

> Those masterful images, because complete
> Grew in pure mind, but out of what began?
> A mound of refuse or the sweepings of a street,
> Old kettles, old bottles, and a broken can,
> Old iron, old bones, old rags, that raving slut
> Who keeps the till. Now that my ladder's gone,
> I must lie down where all the ladders start,
> In the foul rag-and-bone shop of the heart.

How is it possible, though, to revoke 'masterful images' in images that are themselves masterful? Can one renounce 'completion' with epithets and rhyme-patterns that in themselves retain a certain repleteness? T. S. Eliot's 'Marina' has been described as a 'poem that stammers into the hardly sayable' but I do not understand this remark. Though Eliot advocates humility and surrender, I do not think that he ever consciously surrenders rhetorical mastery. 'And why should he?' would be a fair question; but if I observe that 'Marina' seems to me to be an extremely eloquent poem and eminently 'sayable', I do so in the context of that obsessive self-critical

Romantic monologue in which eloquence and guilt are intertwined, and for which the appropriate epigraph would be one abrupt entry in Coleridge's 1796 Notebook: 'Poetry – excites us to artificial feelings – makes us callous to real ones'.

There is a striking paragraph in Hannah Arendt's essay on Walter Benjamin in which she argues that to describe him and his work at all adequately 'one would have to make a great many negative statements'; as, for example, 'his erudition was great, but he was no scholar . . . he was greatly attracted . . . by theology . . . but he was no theologian and he was not particularly interested in the Bible . . . he thought poetically, but he was neither a poet nor a philosopher'. This quotation is central, indeed crucial, to the presentation of my argument. I have already conceded that, however challenging my title may appear, it nonetheless conforms. It is a not unfamiliar modernist theory which 'requires art to be destructive', which 'takes the violence of novelty as essential to success'. I may choose to ignore this theory, but I can't seem to be ignorant of it. I have to say, therefore, that the 'menace' to which I propose to refer is not that species of anti-bourgeois terrorism with which the names of Baudelaire, the Surrealists and Antonin Artaud have been indiscriminately linked. Nor is it the menace of the poetry of Négritude as polemically invoked by Sartre in writing of Césaire: 'Surrealism, a European poetic movement, is stolen from the Europeans by a Negro who turns it against them'. Nor is it that menace to which Hugh Kenner alludes in his epitaph for the American poet H. D.: 'Her grown life was a series of self-destructions, her poetic discipline one of these'.

As for 'atonement', the modern age is not unfamiliar with a literature of penitence; there is even, one may add, a literature of penitential literature. Thomas Mann is on record as saying that his novel *Doctor Faustus* is 'confession and sacrifice through and through'. The sin of Mann's protagonist, the composer Adrian Leverkühn, is in some respects similar to that which Maritain termed '"angelism," the refusal of the creature to submit to or be ruled by any of the exigencies of the created natural order'. And yet, of course, such a refusal to submit to these exigencies has itself been seen as the crime of capitalism, imperialism, modern technology and technological warfare. So there is a sense in which the modern artist is called upon to atone for his own illiberal pride and a sense in which he is engaged in vicarious expiation for the pride of the culture which itself rejects him. He can't win; but, you might say, he can't lose either; for in the words of Grotowski, in his book *Towards a Poor Theatre*, the actor 'does not sell his body but sacrifices it. He repeats the atonement; he is close to [secular] holiness'. It is, you may well feel, the sort of testimonial at which one looks twice.

Hannah Arendt, on the other hand, is reluctant to 'recommend' Walter

Benjamin to our attention, to adjust his solitary witness to any of our recommended categories. One respects her scruple and her strategy. The 'negative statements' through which she vindicates, against the current of assumption, the man she believes Benjamin to be, themselves constitute a form of Romantic mimesis. Readers of the *Biographia Literaria* may note that Coleridge's concern is not so much with thought as with 'the mind's self-experience in the act of thinking' and that this 'self-experience' is most clearly realized by the process of 'win[ning one's] way up against the stream' or of observing how 'human nature itself [fights] up against [the] wilful resignation of intellect' to the dominion of common assumption and mechanical categorization. For Matthew Arnold, in his essay 'The Function of Criticism at the Present Time', the crucial vindication of Burke's integrity is his capacity to 'return . . . upon himself'; and a recent critic has described the Odes of Keats in precisely these terms: 'There was for Keats a certain justness, perhaps even a necessity, in beginning the first of the Odes [of 1819: the 'Ode to Psyche'] by a return upon himself'. It is, of course, a frequently observed fact that the first word of the final stanza of Keats's 'Ode to a Nightingale' ('Forlorn! the very word is like a bell') echoes the last word of the preceding stanza ('Of perilous seas, in faery lands forlorn'). The echo is not so much a recollection as a revocation; and what is revoked is an attitude towards art and within art. The menace that is flinched from is certainly mortality ('Where youth grows pale, and spectre-thin, and dies') but it is also the menace of the high claims of poetry itself. 'Faery lands forlorn' reads like an exquisite pastiche of a Miltonic cadence: 'Stygian cave forlorn' ('L'Allegro', 1.3); 'these wilde Woods forlorn' (*Paradise Lost*, IX, 910). We perhaps too readily assume that the characteristic Romantic mode is an expansive gesture ('Hail to thee, blithe Spirit! Bird thou never wert'). That which MacKinnon has described, in speaking of Kant, as a 'tortuous and strenuous argument, whose structure torments the reader' is equally a paradigm of Romantic-Modernist method:

> Not, I'll not, carrion comfort, Despair, not feast on thee;
> Not untwist – slack they may be – these last strands of man
> In me or, most weary, cry *I can no more.* I can;
> Can something, hope, wish day come, not choose not to be.

These lines by Hopkins may also be said to embody the positive virtue of negative statements, which I have already remarked in Hannah Arendt's essay.

As I have also previously remarked, we are not unfamiliar with a modern literature of penitence, nor indeed with that required secondary reading which is at times, and not inappropriately, a penance in itself. One is, so to speak, 'winning one's way up against the stream'. In the past twenty years or so, in both Europe and North America, there has been a proliferation of studies devoted to aspects of the inter-relationship of theology and literature, or 'the coinherence of religion and culture', to use the wider terms which some prefer. Professor Nathan A. Scott and his colleagues in the Divinity School of the University of Chicago have produced, and have inspired others to produce, a considerable body of criticism and exegesis which may fairly be described as being ecumenically nourished by the work of the Catholic Maritain and the Protestants Tillich and Bonhoeffer. It would seem unreasonable not to concur with Scott's précis of the situation; his suggestion that among 'the principal motives that underlie the general movement of [literary] criticism in our period' is the attempt 'to offer some resistance to the reductionist tendency of modern scientism', and that such 'resistance' is vulnerable to its own reductionist tendency whereby the precious autonomy of the poem may appear as no more than a structure of grammar and syntax. Scott is opposed to both 'reductionist scientism' and aesthetic hermeticism. He quotes with approval Vivas' view of the aesthetic experience as 'intransitive attention' and de Rougemont's definition of the work of art as 'a calculated trap for meditation'. The key terms for Scott's argument are 'attention' and 'meditation' because both words suggest not only an active contemplation of minute particulars and a resistance to sentimental substitution but also an 'ultimate concern' for 'the world of existence that transcends the work'. My reservations, I have already implied, relate not so much to principle as to practice. Although he recognizes that the artist only 'makes good his vocational claim . . . by the extent of the success with which he shapes the substance of experience', Scott's interpretation of what constitutes 'substance of experience' strikes me as being more simply discursive and more tenuous than his endorsement of such terms as 'intransitive attention' leads one to expect. Having been told, in the course of a single essay on Saul Bellow, that *Henderson the Rain King* is an 'adventure in atonement', that 'the comedy of *Herzog* is a comedy of redemption', that 'the drama [in the late books] becomes explicitly a drama of reconciliation', it is with a peculiar urgency of assent that one recalls MacKinnon's remark: 'the language of repentance is not a kind of bubble on the surface of things' or encounters Henry Rago's proper insistence that 'when the language is that of the imagination, we can be grateful enough to read that language as it asks to be read: in the very density of the medium, without the violence of interpolation or reduction'.

It is, I think, crucial at this point to draw a distinction between, on the one hand, a formal acknowledgement of the human condition of anxiety or guilt and, on the other, 'the empirical guilty conscience'. It is one thing to talk of literature as a medium through which we convey our awareness, or indeed

6

our conviction, of an inveterate human condition of guilt or anxiety; it is another to be possessed by a sense of language itself as a manifestation of empirical guilt. In G. K. Chesterton's study, *Charles Dickens*, he remarks that 'a saint after repentance will forgive himself for a sin; a man about town will never forgive himself for a *faux pas*. There are ways of getting absolved for murder; there are no ways of getting absolved for upsetting the soup'. In Helen Waddell's novel *Peter Abelard*, we encounter the thought again, shorn of its risible bathos and delivered with a becoming genial ironic *gravitas* by Gilles de Vannes, Canon of Notre Dame:

> For one can repent and be absolved of a sin, but there is no canonical repentance for a mistake.

Out of context this has just the right weight and edge to enhance the thesis and the occasion. But it may be that Chesterton grasped the truth of the matter. Under scrutiny, this is the essence to which my term 'empirical guilt' is reduced: to an anxiety about *faux pas*, the perpetration of 'howlers', grammatical solecisms, misstatements of fact, misquotations, improper attributions. It is an anxiety only transiently appeased by the thought that misquotation may be a form of re-creation. This thought, originally proposed by Dr Matthew Hodgart in 1953, is subjected to intense and challenging scrutiny by Professor Christopher Ricks in an article in *The Times Literary Supplement*. Ricks's essay vindicates one's anxiety but nothing relieves it. Well, if one feels like this about it, why carry on? And why carry on so? And in public too!

'No man but a blockhead ever wrote, except for money.' Like Boswell, I feel a little distress at the good Doctor's blunt remark and like Boswell I hasten to add that 'numerous instances to refute this will occur to all who are versed in the history of literature'. If, however, we choose to take Johnson's words as a figure of speech implying that all men write from impure motives, whether from that 'necessity' which he himself cites, or from a desire for 'wreaths of fame and interest', or for 'erotic honey', or whatever, Johnson's cynicism may seem more sustainable. If that is so, let us postulate yet another impure motive, remorse, and let us suggest that a man may continue to write and to publish in a vain and self-defeating effort to appease his own sense of empirical guilt. It is ludicrous, of course. 'A knitting editor once said "if I make a mistake there are jerseys all over England with one arm longer than the other".' Set that beside Nadezhda Mandelstam's account of the life and death of her husband, the Russian poet Osip Mandelstam, and one can scarcely hope to be taken seriously. Men are imprisoned and tortured and executed for the strength of their beliefs and their ideas, not for upsetting the soup. And yet one must, however barely, hope to be taken

7

seriously. It seems to me one of the indubitable signs of Simone Weil's greatness as an ethical writer that she associates the act of writing not with a generalized awareness of sin but with specific crime, and proposes a system whereby 'anybody, no matter who, discovering an avoidable error in a printed text or radio broadcast, would be entitled to bring an action before [special] courts' empowered to condemn a convicted offender to prison or hard labour. It may well strike others as unassailable evidence that the woman was merely an obsessional neurotic. Perhaps one could phrase the matter more moderately and say that one does not regard it as at all eccentric to endorse the view that grammar is a 'social and public institution', or to share W. K. Wimsatt's belief in 'the fullness of [the poet's] responsibility as public performer in a complex and treacherous medium'.

Stephen Spender, in his useful little book on Eliot, raises the question of the distinction between 'legal crime' and 'sin' in Eliot's thought. In *The Elder Statesman*, Lord Claverton declares:

> It's harder to confess the sin that no one believes in
> Than the crime that everyone can appreciate.
> For the crime is in relation to the law
> And the sin is in relation to the sinner.

Spender says that 'the point Eliot is trying to make is of course that "sin" is worse than "crime"'. The logic, if it is logic, underlying Claverton's words is that it is made more objectively difficult to confess if no one apart from oneself believes that there is anything which needs to be confessed. Spender is sceptical of the manner in which Eliot demonstrates his distinction and his priorities; and I would agree that some scepticism is justified. In *The Family Reunion* and *The Elder Statesman* 'sin' is more important than 'crime' partly because the criminal act is ultimately revealed to be either non-existent or very much less than one had been led to imagine. Reflecting upon his last play *The Elder Statesman*, one is inclined to wonder how far Eliot has succeeded in distinguishing 'sin' from those other mental or psychic states which solipsists might confuse with it. The late Harry Guntrip once suggested:

> It may be that the practical and relevant approach to the problem of sin for this age is by the study of the devastations, personal, social, and spiritual, which are the product of anxiety.

Doubtless he has a very strong case; but to Eliot, despite the portrayal of 'pathological despondency', 'psychic impotence', in *The Waste Land* and other poems, it would seem, possibly, a blurring of categories, an abdication

8

of priorities. Yet that which Claverton retrospectively describes, and what he immediately undergoes when accosted by Gomez and Mrs Carghill, seems very like one aspect of that condition which Guntrip calls 'anxiety'. But that is precisely Eliot's point, you will fairly remonstrate. In a secular age we experience anxiety until we learn to read ourselves aright and know that we act and suffer as creatures of sin. Even so, I still maintain that something has eluded Eliot, eluded him in 'the very density of the medium'. Grover Smith, in his essay on *The Elder Statesman*, says that 'Claverton, troubled by his role in [the past lives of Gomez and Mrs Carghill], is indifferent to their future, though neither has wronged him so much as he has wronged them. He makes no atoning gesture'; but Smith seems to imply that this is an ironic profundity of Eliot's making. It seems to me, however, that in determining the order of priority between 'sin' and 'anxiety' 'the kind of pleasure that poetry gives' is to be experienced through contact with the force-fields of these conflicting yet colluding entities. To control such forces demands an *askesis* rather different from the 'ascetic rule' which Eliot laid down for himself in the writing of dramatic verse:

> the ascetic rule to avoid poetry which could not stand the test of strict dramatic utility.

That 'poetry' which is excluded on utilitarian grounds is, I would argue, that very element which could master the violence of the conflict and collusion between the sacramental and the secular, between the dogmatic exclusiveness of 'sin' and the rich solipsistic possibilities of 'anxiety'. I would further suggest that Eliot's asceticism in the three post-war verse-plays is too often a kind of resignation, or what W. W. Robson, in an acute criticism of the later essays, calls 'abstention'. He is left with a language that is at once aloof and ingratiating, unambiguous yet ambivalent. In the essay 'Poetry and Drama' Eliot speaks of 'a fringe of indefinite extent, of feeling which we can only detect, so to speak, out of the corner of the eye and can never completely focus . . . At such moments, we touch the border of those feelings which only music can express'. As Eliot well knew, however, a poet must also turn back, with whatever weariness, disgust, love barely distinguishable from hate, to confront 'the indefinite extent' of language itself and seek his 'focus' there. In certain contexts the expansive, outward gesture towards the condition of music is a helpless gesture of surrender, oddly analogous to that stylish aesthetic of despair, that desire for the ultimate integrity of silence, to which so much eloquence has been so frequently and indefatigably devoted.

Edward Mendelson, the editor of the posthumous *Collected Poems* of W. H. Auden, has said that 'as he grew older Auden became increasingly distrustful of vivid assertions, increasingly determined to write poems that

9

were not breathtaking but truthtelling' and has endorsed the poet's motives and actions with his own suggestion that 'the local vividness of a line or passage can blind a reader into missing a poem's overall shape'. I would suggest, however, that the proof of a poet's craft is precisely the ability to affect an at-one-ment between the 'local vividness' and the 'overall shape', and that this is his truthtelling. When the poem 'comes right with a click like a closing box', what is there effected is the atonement of aesthetics with rectitude of judgment. The suggestion that the proof of a poet's integrity is a conviction that he must sacrifice 'vividness' to 'shape' seems to me to stem from a very dubious philosophy of authorial responsibility to the 'reader'. My argument is thus obliged to distinguish between this matter of 'empirical guilt' which is involved with 'the density of the medium' and the principles of Christian penitence and humility which were, it seems reasonable to suggest, the disciplines of conscience within which Eliot and Auden increasingly worked. One is left with the awkward observation that the acceptance of a principle of penitential humility in the conduct of life does not necessarily inhibit a readiness to accept the status of 'maestro' conferred by a supportive yet coercive public. It's worse than awkward, it's damned awkward; it cannot but be seen as a churlish refusal to concede honour where honour is due. I would reply that it is not a matter of *ad hominem* rebuke but a suggestion that fashionable adulation of the 'maestro' when there is so little recognition of the 'fabbro', 'homo faber', is one aspect of what C. K. Stead mordantly but not unfairly calls the 'struggle between poets and "poetry-lovers"', except that the very word 'struggle' suggests purpose and engagement. As Jon Silkin has remarked, 'it is not disagreement we have now but deafness'. Deafness, yes; and arbitrary assumption. To 'assume' is literally 'to take to oneself, adopt, usurp'; and the fashion in which society can 'take up' and 'drop' the poet (as John Clare was taken up, and dropped) is a form of usurpation which has little or no connection with intrinsic value. 'Where are our war poets?', and all that.

In April 1915, six months before he himself was killed in action, aged twenty, the poet Charles Hamilton Sorley wrote, in a letter home:

> [Rupert Brooke] is far too obsessed with his own sacrifice, regarding the going to war of himself (and others) as a highly intense, remarkable and sacrificial exploit, whereas it is merely the conduct demanded of him (and others) by the turn of circumstances, where non-compliance with this demand would have made life intolerable . . . He has clothed his attitude in fine words: but he has taken the sentimental attitude.

Sorley's criticism of Brooke is at once brilliant in its economy and far-reaching in its adumbration. I even suggest that it atones for Brooke's 'sentimental attitude'; that is, it brings together details and perceptions which the 'sentimental attitude' has arbitrarily set apart and at odds. When Sorley turns upon the nub of the question it is with a turn of phrase precisely evoking the supportive yet coercive role of militant cliché. Two days before Sorley wrote this letter, Churchill's obituary on Brooke had appeared in *The Times*:

> A voice had become audible, a note had been struck, more true, more thrilling, more able to do justice to the nobility of our youth in arms engaged in this present war, than any other . . .

If ever anyone died in the nobility of his youth it was Charles Sorley, 'shot in the head by a sniper, as he led his company at the "hair-pin" trench near Hulluch'. Within the ambience of Churchill's rhetoric it was Sorley rather than Brooke who died the exemplary death. It is, of course, open to suggestion that, in so dying, he abjured the witness of his own poetic intelligence, the intelligence that could so illuminate his brief but trenchant comment on Brooke's lyrical 'compliance with a demand'; and that, in so doing, he became a passive accessory to rampantly jingoistic poesy. But I see it differently. It is not that public rhetoric degrades Sorley's acute perception but that Sorley's perceptive statement redresses and redeems the rhetoric of the *Times* obituary and establishes a sounder basis for judging the nature of exemplary conduct.

It may well seem that, at this point, my argument vacillates. Having persistently stressed that we encounter both the 'menace' and the atoning power of poetry within the 'indefinite extent' of language itself, I now make my appeal to something else. In seeking to present Sorley as an atoning agent, I support my appeal to the 'poet's intelligence' with a gesture towards something beyond words. To this objection I would respond that 'utterance' and 'act' are not distinct entities. A statement by the philosopher Rush Rhees gives one some ground for such a belief:

> For we speak as others have spoken before us. And a sense of language is also a feeling for ways of living that have meant something.

Sorley, we may say, 'saw it through' even as he 'saw through it'; and I cannot believe that this could be regarded as the abjuring of 'thought' by 'deed'. I can see it only as an exemplary instance of the at-one-ment of the 'sense of language' with the feeling for the ways of life.

Charles Sorley's brief life was remarkably 'at unity with itself'; and one must inevitably ask to what extent 'exemplary atonement' is possible when the life is otherwise. A student of Coleridge has recently written:

11

it is precisely through [the] perpetual dialectic between ideal and sinful reality that Coleridge is able to introduce his particular fusion of intellectual and *moral* qualities. In his experience as a ruined man, moral awareness, and even wisdom, were encountered through failure rather than success.

And one would add that what is especially noteworthy is the quality of disinterested stoicism with which this habitually self-pitying man was able to bring his own broken life and aspirations into the focus of meditation.

> In silence listening, like a devout child,
> My soul lay passive, by thy various strain
> Driven as in surges now beneath the stars,
> With momentary stars of my own birth,
> Fair constellated foam, still darting off
> Into the darkness; now a tranquil sea,
> Outspread and bright, yet swelling to the moon.

That is from the poem 'To William Wordsworth: composed on the night after his recitation of a poem on the growth of an individual mind' (1807). The beauty of the image of 'fair constellated foam' does not conceal the nature of the experience which Coleridge is evoking. The brightness is ephemeral, it moves outward from the centre into the darkness where it is quenched or lost. But we note also that there are other stars; and, bearing in mind the broodingly complex nature of Coleridge's inspiration, it may be legitimate to relate the first reference to the 'stars' to those of the prose gloss (virtually a marginal prose poem) added to the 1817 edition of 'The Ancient Mariner': in particular to a passage which Humphry House rightly calls 'that one long sentence of astounding beauty' and which I would call an outstandingly beautiful image of the attainability of atonement:

> In his loneliness and fixedness he yearneth towards the journeying Moon, and the stars that still sojourn, yet still move onward; and every where the blue sky belongs to them, and is their appointed rest, and their native country and their own natural homes, which they enter unannounced, as lords that are certainly expected and yet there is a silent joy at their arrival.

Flux is redrawn as harmonious motion towards rest; and rest is seen as active contemplation, not as stagnation. The mariner knows enough of stagnation; the stars are of another order. And in the poem 'To William Wordsworth', the private utterance of highly organized art can for a while stabilize the self-dissipating brilliance of the listener's mind, that is, Coleridge's mind, the mind that is concentrating upon that very diffusion. It is a transfiguring of weakness into strength, a subsuming, which Kathleen

12

Coburn effectively characterizes in a note concerning the title of her 1973 Riddell Lectures on Coleridge, *The Self Conscious Imagination*:

> The two senses of the word are thus antithetical, *self-conscious* (1) as being realistically accurate about one's identity, and *self-conscious* (2) as being anything but clear, in fact painfully in doubt.

One must try to balance the accounts. I have argued, with MacKinnon, that 'the language of repentance is not a kind of bubble on the surface of things' and I have endorsed Rago's suggestion that we comprehend such language 'in the very density of the medium'. But from that point I sustain two views of the matter, divergent views, you may well think, which threaten to tear my argument apart. In speaking of Charles Sorley, and the *Times* obituary on Brooke, I have seriously proposed that one writer can make vicarious atonement for other writers' sins of commission and omission. In speaking of Coleridge's poem 'To William Wordsworth' I have suggested that a poet can transfigure his own dissipation by a metaphor that perfectly comprehends it. But I have added a comic sub-plot, this business of 'empirical guilt' which, as I seem to have conceded, appears to be nothing more than a trivial bourgeois anxiety about upsetting the soup. What credence can be given to remorse over a *faux pas* in a world of actual terror, in which poetry is increasingly asked to be 'prophetic, menacing, terrorist, violent, protesting'?

In his diary for the years 1935–50, *Il Mestiere di Vivere* (translated into English as *This Business of Living*), Cesare Pavese comments that 'the political body does not die and so does not have to answer for itself before any God'. It is the entry for 14 June 1940. Was Pavese, perhaps, aware of Benedetto Croce's comment on Machiavelli's *Il Principe*: that it shows 'a clear recognition of the necessity and autonomy of politics, of politics which is beyond, or rather, below moral good and evil, of politics which has its own laws against which it is useless to rebel'? In the second of his 1969 T. S. Eliot Memorial Lectures, Conor Cruise O'Brien gives a sharply critical account of those whom he calls 'American neo-Burkeans' and their belief that Burke's central achievement was 'to close the Machiavellian schism between politics and morality'. O'Brien considers this belief to be false. In his view 'the famous schism was not closed at all but was simply one of those distressing matters, abounding in the Burkean universe, for which some arrangement of veils was normally appropriate'. If O'Brien is right, and I cannot see that he is wrong, he is nonetheless correct to concede, in the same lecture, that Burke 'understood the density of the human medium in which change occurs, and the cost of change'. I do not overlook my own caveat about the need to resist the attraction of terminology itself, when I remark the affinity of phrase that connects O'Brien's allusion to the nature of society ('the density of the human medium'), Wimsatt's reference to 'a complex and treacherous medium' and Rago's suggestion that we read the language of the imagination 'as it asks to be read: in the very density of the medium'. You may say that I am merely proving susceptible to the influence of random association; but in answer I would say that these three 'densities' are more than mutually attractive correlatives. One is involved with something other than a 'conceptual elaboration of the similarity between literary and moral judgment'. It is rather a recognition that in the act of 'making' we are necessarily delivered up to judgment. Among contemporary theologians it is D. M. MacKinnon who, unless I grievously misread him, most acutely perceives and articulates the matter which I am here struggling to express. In the course of a discussion of Butler's ethics, he refers to a suggestion which, as he says, perhaps 'strikes us as ill-conceived and old-fashioned . . .' the 'suggestion of an "ought" somehow imposing itself upon us out of the matter of the actual'. I hope it will not travesty the significance of this sentence if I say that, to my mind, it is a way of closing the 'schism' not only between politics and morality but also between 'literary and moral judgment', a 'schism' which the very word 'similarity' only serves to emphasize. In the light of these observations I am bound to say that O'Brien's dictum, 'the study of literature is a social science', does not adequately take the strain of his own argument. That 'is' is itself schismatic. The cogency of his criticism is perhaps more truly appreciated if we sense a continuity between the Coleridgean 'self-conscious imagination' and O'Brien's own 'suspecting glance'. If the socio-political 'scene' in recent years has been characterized by an unsuspecting allegiance to 'slogans [and] sages', by the worship of charisma, instant wizardry and all that is 'technically sweet', we may ask to what extent literary aesthetics have colluded with such sentimentality and cynicism. I consider it at least open to argument that, in such an epoch, the sense of 'empirical guilt' involved with what can be termed 'culpably careless proof-reading' has an intrinsic value, for, in such a context (to quote MacKinnon once more), 'one can never be quite at ease in the presence of the suggestion that . . . a teleological ethic need not have the slightest truck with utilitarianism'.

It is perhaps fitting that a debate such as this should convey an apprehension of its own trespass; a trespass inextricably involved with tradition and decorum. In speaking on such a topic I am a mere stumbling latecomer to a field, or fields, already traversed by distinguished scholars closely associated, at some period of their work, with the University of Leeds: J. M. Cameron, Fr Martin Jarrett-Kerr, Alasdair MacIntyre, Walter Stein, the late Harry Guntrip. I would be loath to embarrass any of these with the burden of my discipleship. In certain circumstances it is possibly more courteous, as well as scrupulous, to be taken for an apostate rather than a disciple. I also sense that I have made myself vulnerable to accusations of having strayed upon domains of knowledge which I have no right to 'profess', where I can at best speculate like an amateur without credentials. Auden remarked that 'it is both the glory and the shame of poetry that its medium is not its private property'. Philosopher, theologian, historian and sociologist are pleased to find in literature both quarry and common ground. Excursions in the opposite direction tend to be regarded with more suspicion. Let me say only that I am trying to follow a thread; I am not attempting to trample a right-of-way.

I may also be accused of wandering even within my own proper domain. My theme is 'Poetry as "Menace" and "Atonement"', yet I have not restricted my examples to works written in rhyme and metre. Little apology is needed here, I think, 'since there have bene many most excellent Poets that never versefied, and now swarme many versefiers that need never answere to the name of Poets'. Even so, my argument puts a very particular pressure upon that which I elect to call 'poetry'. Claude Lévi-Strauss has said that 'the poet behaves with regard to language like an engineer trying to form heavier atoms from lighter ones'. Karl Barth remarked that Sin is the 'specific gravity of human nature as such'. I am suggesting that it is at the heart of this 'heaviness' that poetry must do its atoning work, this heaviness which is simultaneously the 'density' of language and the 'specific gravity of human nature'. There is perhaps no need for me to point out that my thesis is as much symptomatic as diagnostic, that in its account of certain aspects and effects of Romanticism it is itself a part of that which it describes; in some respects its tendency to 'swim up against the stream' of much current thinking about the nature and function of poetry is itself a minor Romantic trait. The major Romanticism of our time, or that which some propound as the major Romanticism, sees the poet's vocation as a 'searching for a way of reconciling human vision with the energies, powers, presences, of the non-human cosmos'. Charles Olson has described the poem as a 'high energy-construct and, at all points, an energy-discharge'. In such cases the 'menace' of poetry may be taken as referring not only to the 'energy' which is to be released, at whatever cost, but also to the inevitable fatalities occurring in any high-risk occupation. In my thesis, however, the idea of 'menace' is entirely devoid of sublimity: it is meanly experiential rather than grandly mythical. The poet as I envisage him is quite unlike the Baudelaire of Eliot's celebrated panegyric ('Baudelaire was man enough for damnation'); whereas the craft of poetry itself, as I describe it, comes close to resembling that 'frightful discovery of morality' to which Eliot alludes in one of his finest passages, the account of the nature of Beatrice in Middleton's and Rowley's play *The Changeling*:

15

In every age and in every civilization there are instances of the same thing: the unmoral nature, suddenly trapped in the inexorable toils of morality – of morality not made by man but by Nature – and forced to take the consequences of an act which it had planned light-heartedly. Beatrice is not a moral creature; she becomes moral only by becoming damned.

But even though I choose to regard that vision of 'the unmoral nature suddenly trapped in the inexorable toils of morality' as an oblique yet penetrating insight into the nature of the creative act, the resemblance is imperfect; and in that imperfection lies our ambiguous hope. The reason why the poet's 'discovery' is finally not to be confused with Beatrice's 'discovery' is perhaps implied by my reference to that *vision* of "the unmoral nature ..."' by which I attempt to set at one the piercing insight and the carnal blundering, in which I intentionally recollect Coleridge's capacity to 'transfigure his own dissipation by a metaphor that perfectly comprehends it'. It has been said that 'real poets often predict their own future in verse'. One may choose to glean from this what optimism one can. Julian Green is 'somewhat of the opinion' that the poet Charles Péguy 'was providentially influenced by his own work' and H. A. Williams has said 'the academic study of prayer may lead a man to pray'. Or one may choose to read it in a darker light. The seventeenth-century divine, Jeremy Taylor, drew a significant distinction between 'attrition' and 'contrition'. 'Attrition begins with fear, Contrition hath hope and love in it; the first is a good beginning, but it is no more'. And Fr Jarrett-Kerr writes that 'repentance is an attitude of mind which implies readiness to have the mind changed (*meta-noia*)'. It is therefore conceivable that a man could refuse to accept the evident signs of grace in his own work; that he himself could never move beyond that 'sorrow not mingled with the love of God' even though his own poems might speak to others with the voice of hope and love.

If only as a formality one should perhaps make explicit what has been implicit throughout this discussion. It is evident that my argument is attracted, almost despite itself, towards an idea by which it would much prefer to be repelled. But surely, one may be asked to concede, it is more than attraction. Is it not a passionate adherence; a positive identification with the magnificent agnostic faith whose summation is in the 'Adagia' of Wallace Stevens? –

After one has abandoned a belief in god [*sic*], poetry is that essence which takes its place as life's redemption.

Stevens is here in the tradition of Arnold and the Symbolists. Vincent Buckley has suggested that Arnold 'explicitly associates the notion of mastering a hostile world with the interpretative power of poetry' and Arthur Symons in *The Symbolist Movement in Literature*, a minor work which had not a little influence on Eliot and Yeats, celebrates the making of poetry as a sacred task. We may note how persistently Symons's book dwells upon the concept of 'mastery', whether 'laboriously acquired', as with Huysmans, practised with 'supreme disinterestedness' by Mallarmé, or sensually-instinctual in the genius of Rimbaud. According to Symons, Rimbaud 'brought into French verse something of that "gipsy way of going with nature, as with a woman"; a very young, very crude, very defiant and sometimes very masterly sense of just those real things which are too close to us to be seen by most people with any clearness.' The major caveat which I would enter against a theological view of literature is that, too often, it is not theology at all, but merely a restatement of the neo-Symbolist mystique celebrating verbal mastery; an expansive gesture conveying the broad sense that Joyce's *Ulysses* or Rilke's *Duino Elegies* 'must, in the splendor of its art, evoke astonishment at the sheer magnificence of its lordship over language'. If an argument for the theological interpretation of literature is to be sustained, it needs other sustenance than this.

P. T. Forsyth writes that 'the effect on us of the moral ideal is not simply admiration; it is confusion; it is accusation; it is judgment . . . Its very grandeur fills us with a sense of weakness, nay, of blame, shame and despair'. And yet through this very blame, shame and despair, we rise to a discovery of true personality. 'The man who does not rise to be a person', says Forsyth, 'becomes an item . . . He knows nothing of action, only of incident . . . He knows nothing of responsibility, of guilt, of sin; and his only goodness is goodness of heart, because he is built that way, and that is the way of least resistance, and is always popular'. It seems not unreasonable to draw from these two statements one's own conclusion: that in the constraint of shame the poet is free to discover both the 'menace' and the atoning power of his own art. However much and however rightly we protest against the vanity of supposing it to be merely the 'spontaneous overflow of powerful feelings', poetic utterance is nonetheless an utterance of the self, the self demanding to be loved, demanding love in the form of recognition and 'absolution'. The poet is perhaps the first to be dismayed by such a discovery and to seek the conversion of his 'daemon' to a belief in altruistic responsibility. But this dismay is as nothing compared to the shocking encounter with 'empirical guilt', not as a manageable hypothesis, but as irredeemable error in the very substance and texture of his craft and pride. It is here that he knows the affliction of 'being fallen into the "they"' and yet it is here that his selfhood may be made at-one with itself. He may learn to live in his affliction, not with the cynical indifference of the reprobate but with the renewed sense of a vocation: that of necessarily bearing his peculiar

17

unnecessary shame in a world growing ever more shameless. He may 'rise to be a person' in a society of aggregates and items; he may even transfigure and redeem that 'word-helotry' to which Dr George Steiner sees the merely literate man ultimately condemned in a culture divided between electronic data-processing and music. 'Attrition begins with fear'. True; but perhaps, as William Empson has said, in his verse-meditation after Bunyan:

> To take fear as the measure
> May be a measure of self-respect. Indeed
> As the operative clue in seeking treasure
>
> Is normally trivial and the urgent creed
> To balance enough possibles; as both bard
> And hack must blur or peg lest you misread;
>
> As to be hurt is petty, and to be hard
> Stupidity; as the economists raise
> Bafflement to a boast we all take as guard;
>
> As the flat patience of England is a gaze
> Over the drop, and 'high' policy means clinging;
> There is not much else that we dare to praise.

TWO

The Absolute Reasonableness of Robert Southwell

When Robert Southwell wrote of the 'inhuman ferocity' with which Catholic recusants were treated in Elizabethan England he was not toying with hyperbole. 'Grinding in the Mill, being beaten like slaves, and other outragious vsages' were, again in his own words, but 'ordinary punishments'; and from the extraordinary, the 'more fierce and cruell' penal torments inflicted upon certain priests and laymen, our powers of contemplation recoil. Ours, but not his. Southwell's prose writings, with the exception of the brief but crucial *Spiritual Exercises and Devotions*, were not intended primarily for his own ascetic meditational practice; they were, as the title of his major prose work makes explicit, epistles of comfort 'to the Reverend Priestes, & to the Honorable, Worshipful, & other of the Laye sort restrayned in Durance for the Catholicke Fayth'; but they are not without indications that by such means he was able to apply Ignatian practice to a double purpose, 'seeing in imagination the material place where the object is that we wish to contemplate'. The 'object contemplated' was most frequently and formally the Passion of Christ; but there can be little doubt that for Southwell it was also his own 'almost inevitable martyrdom'. The pioneering modern historian of recusancy, A. O. Meyer, remarked that a 'yearning for martyrdom' was 'the only fault that can be found with the priests of the Elizabethan age'; but in adding that 'the death of the martyrs ever remained the catholic mission's most effective means for achieving its purpose' he may be thought to have annulled the force of the criticism and to have provided, albeit obliquely, a means for understanding the quality of mystical pragmatism which illuminates much of Southwell's work. I would justify my use of the dangerous term 'mystical' by referring to J. R. Roberts's statement that, in Southwell's poem, 'The Burning Babe', 'there is a note . . . of his having lost himself in ecstatic delight, the goal of Ignatian methodology', although I have certain reservations about Roberts's emphasis and although I would not wish to claim this as the keynote for Southwell's work, or even as the prime characteristic of the Ignatian method practised by the Jesuit missionaries in Elizabethan England. Southwell, as his letters to Aquaviva, Agazario and Robert Persons reveal, was minutely and meticulously practical in his conduct of missionary matters, seeking permission, for instance, to 'bless 2,000 rosaries and 6,000 grains, for here all are asking for such objects'. Southwell's predecessor in the field, Edmund Campion, had 'particularly recommended . . . that such of the Society as should be sent upon the *English* Mission, should be able Preachers' and Southwell's own suggestions were fully in accord with this precedent. He wrote to Fr Agazario in Rome: 'Every priest here is useful, especially those who are well skilled in moral theology and controversy . . . Preachers are here in great request: hence it is most important that the students should practise themselves, so as to acquire readiness of speech and a plentiful supply of matter'. As the editors of the Clarendon Press text of his poems remind us, 'English was for him the language of his apostolate'. After years of exile he needed to apply himself 'with much diligence to the study of his native tongue'.

Southwell does not shirk the word 'controversy'; and neither should we. Louis L. Martz has written of 'an England shaken by a threefold controversy, Catholic against Anglican against Puritan'; but Meyer refers to 'the internal divisions within the catholic camp'; and more recent studies confirm this view. Dr John Bossy has drawn attention to the 'mutual recrimination' and 'stress between [recusant] clergy and laity'; and his assessment is endorsed by J. C. H. Aveling's suggestion that 'amongst English Catholics' in general there existed an 'anti-Jesuit opposition of a powerful and virulent kind'. 'Threefold' must therefore become 'fourfold' and the statement rephrased to read 'Catholic against Catholic against Anglican against Puritan'.

Such statements must, at the least, be allowed as caveats against the 'special pleading' of the 'English Catholic legend', as Aveling calls it. Even Pierre Janelle, in his justly esteemed study *Robert Southwell the Writer* (1935), associates certain admirable qualities – 'a cheerful and loving patience, a gentle and restrained manliness' – too exclusively with pre-Reformation Englishness and with Elizabethan recusant Englishness in particular. That such tones are manifestly present, in the key-works of Edmund Campion and Robert Southwell, one would not for a moment dispute. What is disputable is the contention that these qualities are simply attributable to heredity or national characteristics. Martz is on surer ground in associating that 'mild, moderate, and cheerful temper', which characterizes some of Southwell's finest work, with Franciscan tradition and practice. Janelle himself is aware that Franciscan ascetical writings were 'in great favour among the early Jesuits. Southwell was at one with his order in this respect'. If we adopt Janelle's terms, we might call 'a cheerful and loving patience' Franciscan, a 'gentle and restrained manliness' Ignatian; and say that both strands are united in the prose and verse of Southwell and in

Campion's brief masterpiece, his 'Challenge' or 'Letter to the Council', vulgarly known as 'Campion's Brag'.

Southwell composed his *Humble Supplication* 'rapidly and vehemently'; Campion's 'Challenge' was 'written without preparation, and in the hurry of a journey', but one would hesitate to call such works 'spontaneous effusions'. They have nothing in common with that facile 'self-expression' which so debases the current acceptance of 'spontaneity'. Such ease and rapidity as they manifest are the issue of years of arduous rhetorical and meditational discipline, both classical and Ignatian. They are, moreover, the fruit of a 'well-ordered will'; impulse and effect are at one. Fr Christopher Devlin has written that 'the deliberate refusal to allow desire and choice to be separated was the main inspiration of seventeenth-century religious art and poetry' and his words might be justly applied to the work of these two sixteenth-century Jesuits. 'Cheerful . . . patience' and 'restrained manliness' have to be seen as, at one and the same time, the expression of 'desire' and the choice of a 'suitable controversial means of expression'.

The Proclamation of 1591, 'A declaration of great troubles pretended against the Realme by a number of Seminarie Priests and Iesuists . . .', demonstrated the prescience of Campion's 'Challenge', written some ten years earlier, before the full spate of persecutions, in anticipation of just such accusations and slanders. 'Brag' was itself a term of abuse foisted by the opposition on Campion's apologia. The 1591 Proclamation, to which Southwell made an immediate reply, attacked the missionary priests as 'dissolute yong men' disguised in a variety of 'apparell', 'many as gallants, yea in all colours, and with feathers. . . . and many of them in their behauior as Ruffians . . .' Edmund Campion, after his capture, had been addressed by his Anglican adversaries in public disputation as '*miles gloriosus*', that is, as the strutting and braggart mercenary of Plautine and Renaissance comedy. It is in such a context that one must interpret the truism that 'the most important weapons for mission work were casuistry and controversy'. Southwell's controversialist skill, like Campion's, took effect in appearing non-controversial. They abstained from what Helen C. White has pithily termed 'the creative art of denigration' and practised instead a polemic of rapprochement. It was possible, of course, to be master of both styles. In the preface to the 1585 edition of his *Christian Directorie guiding Men to their Salvation*, Fr Robert Persons pleaded that 'a spirite of contradiction and contention . . . for the most parte hindereth devotion' but between 1592 and 1594 he was directing what have been called 'uncompromising and inflammatory writings', some of them against his co-religionists. It was possible, also, for a particular rhetorical gesture to be the mask of a quite contrary intent. The title-page of the 'Rhemes' New Testament (1582), reads like an epitome of this method. 'Translated faithfvlly into English', the work claims to have been undertaken 'for cleering the Controversies in religion, of these daies'; but in the light of the quotation from St Augustine, which serves as one of its two epigraphs, 'cleering . . . Controversies' means confounding 'Heretikes: vvhose deceites cease not to circumuent and beguile . . . the more negligent persons'. The 'Rhemes' 'annotations' are also thoroughly contentious.

'Controversy', says Meyer, 'had its poets too'. One of these, the 'witty and courageous martyr' Richard Gwynne, wrote a number of so-called 'carols' in Welsh. Of Gwynne's song of triumph at the assassination of the Protestant Prince William of Orange, Fr J. H. Pollen remarks that 'it is plainly wanting both in forbearance and in good feeling'. Meyer calls it 'a terrible example of the lengths to which religious excitement can drive a man' but presents it as 'a solitary case, an ugly discord breaking in upon the pure harmony of the poetry of English catholics'. The 'carol' is in fact far less 'terrible' in tone and content than Meyer suggests and, as an example of controversialist rhetoric, seems scarcely remarkable. Even so, those apologetical arguments which attribute fanatical enthusiasm to none but Puritan extremists and 'caddish' behaviour only to erastian deans and time-serving judges and magistrates are notably unfair; and that vision of the pre-dissolution Church, movingly evoked by a modern Catholic historian ('the intimate religion of the little shrines . . . God's Presence in tranquillity in the fields'), is a beautiful but nostalgic image for which some Englishmen already possibly 'hankered' as early as the mid-sixteenth century, and which perhaps bore, and bears, little resemblance to late medieval and early Tudor reality. As Dr Christopher Haigh has cogently argued, in a closely-documented survey, in pre-Reformation Lancashire 'local communities met together only in the churches' and, 'violence was even more likely [there] than elsewhere'. There is a violence of morbid religious excitement, and Dr Haigh gives instances of this. There is also something that Wordsworth called 'savage torpor', concomitant with 'sluggishness of spirit', 'spissa ignorantia'. Elizabethan and Jacobean Catholicism not only endured both kinds, it also perpetuated them, or sank into them, and it is a fair inference that Southwell recognized that the contest to which he had been summoned was to be fought upon a number of fronts.

Richard Topcliffe, the government's chief pursuivant and inquisitor, was an atrocious psychopath, 'homo sordidissimus', as one missionary priest called him, 'a man most infamous and hateful . . . for his bloody and butcherly mind', according to another. When such a creature calls Southwell's cousin Anthony Copley 'the most desperate youth that liveth', one cannot accept the statement without reserve, as Fr Thurston has equably remarked. There is independent evidence, however, to substantiate Janelle's description of Copley as 'a wayward, hot-headed, uncontrollable scapegrace'. Whatever emotional stability he came to possess seems to have been due to Southwell's influence.

Southwell's great predecessor, too, had known similar young Catholics of 'ardent temperament and keen literary appreciations'. Richard Simpson, in his *Edmund Campion: a Biography* (1867), names the members of an 'association' of 'young gentlemen of great zeal and forwardness in religion' who acted as guides and 'lay assistants' to the first generation of Jesuit missionaries on their journeys around England. Numbered among this ardent and appealing company were Anthony Babington and the young poet Chidiock Tichborne. It is surely not without significance that one of Babington's harshest critics was to be Robert Southwell. In a letter to Aquaviva he wrote of 'that wicked and ill-fated conspiracy, which did to the Catholic cause so great mischief'; and in the *Humble Supplication* he referred, with a more aloof contempt, to 'greene witts ... easily ... ouerwrought by Master Secretaries subtill and sifting witt'.

It must be acknowledged that the 1591 Proclamation, in its mocking use of the word 'feathers', though false in tone is probably correct on a point of fact. Southwell was by no means the only missionary priest to concede, uneasily, that it was frequently necessary to appear in 'apparell' quite at odds with 'the graue attire that best suteth our Calling'. To the reasonable question how a turn of phrase can be at once false and true, there is the answer that it is a matter of context; that equity requires a respect for context. It was one thing for Fr Gerard, in hourly peril of arrest, torture and a protracted death, to go about 'garnished with gold or silver lace, satin doublets, and velvet hose of all colours'; it was quite another for the not unlikeable young Anthony Copley, when a student at the English College in Rome, to appear in the pulpit with a rose between his teeth. But slander, no respecter of persons or contexts, would have little difficulty in seeming to equate the one with the other; and in making both appear as gestures of braggart panache, the preening of a *miles gloriosus*, a self-vaunting *os impudens*.

Southwell's necessary choice, therefore, was the achievement of a style ardent yet equable, eloquent and assured yet without 'panache', that is, not tricked out with 'feathers'. It is worth saying again that he was nonetheless a controversialist for sounding non-controversial. Dr William Allen, founder and president of the English College at Douay, had written that, since 'heretics', Protestant controversialists, were wont to 'plume themselves' on their mastery of the vulgar tongue, the future missionaries should practise to 'acquire greater power and grace' in the vernacular. Janelle suggests, a little tendentiously, that in the penultimate chapter of *An Epistle of Comfort* 'Southwell carefully avoids the threatening tone that was so common a feature in the polemical writings of contemporary Puritans'. Ernst Cassirer, it is true, once called puritanism 'a thoroughly quarrelsome and quarrel-seeking religion'; but Christopher Morris fairly reminds us that Archbishop Whitgift's chaplain 'the "saintly" Lancelot Andrewes could make a cruel joke in execrable taste' at the expense of the imprisoned Puritan separatist Henry Barrowe, whose response was 'you speak philosophically but not Christianly'. Catholic and Puritan alike fell victim to Elizabeth's penal laws and Anglicanism was 'as Erastian as Elizabeth herself'. Meyer has remarked that 'no puritan could have surpassed in bitterness and hatred' the vituperation of the 1591 Proclamation whose 'contumelious termes' occasioned from Fr Robert Persons 'a reply full of bitterness, hatred and scorn' and from Fr Robert Southwell *An Humble Supplication*.

In confronting the strategy of the Proclamation Southwell undertook a double defence. Edmund Campion, in his great 'Challenge' of 1580, had appealed for 'fair light', 'good method' and 'plain dealing' to be 'cast upon these controversies, that possibly her zeal of truth and love of her people shall incline her noble Grace to disfavour some proceedings hurtful to the Realm, and procure towards us oppressed more equitie'. The Proclamation lards its threats and vilifications with well-timed cynical gestures of mock reasonableness and tolerance, appealing in the name of all that is 'naturall' and 'honorable' against that which is 'wilfull' and 'monstrous', 'the slanderous speeches and Libelles of the Fugitiues abroad'. In the England of Elizabeth, say the authors of the Proclamation, those 'professing contrary religio[n]' who 'refuse to come to Church' 'are knowen not to be impeached for the same . . . but onely by payment of a pecuniary summe . . .' Answered even in its own terms, without reference to executions and incarcerations, this is the most wilful and monstrous cant. Such travesties of 'fair light', 'good method' and 'plain dealing' are of course far more numbing than the most savage vituperation. Southwell's double defence, I suggest, required him to ensure not only that Campion's pedal-note 'equitie' continued to resonate freely but also that the Proclamation's speciously equable tone was refuted by a 'power and grace' able to judge such a travesty, not so much by attack as by being simply yet manifestly on the level. Southwell appeals, in the manner of Campion, directly to Elizabeth herself, to 'measure [her] Censure with reason and Equity' and to weigh a 'sound beliefe' against a 'shadow of likelihoode':

> For to say we doe [the alleged treasons] vpon hope to be enritched with those possessions that others now enioy hath but very small semblance of probability, considering how much likelyer we are to Inherit your Racks and possesse your places of Execution, then to surviue the present Incumbents of spirituall livings, or live to see any dignities at the King of Spaines disposition.

24

In its poise and resolution, phrase taking the measure of phrase, this passage both seeks and obtains satisfaction for the injury received; but it is a sign of Southwell's unflaunted mastery that the satisfaction does not seem to be the effect of a rhetorical 'turn' or the consummation of a mere *reductio ad absurdum*. We are persuaded that Southwell, in every sense, delivers a just sentence. 'Let it be scanned with Equity, how little seeming of truth [the Proclamation's charge of sedition] carrieth'.

It is noteworthy to what extent Southwell, across the range of the prose and verse composed during his six years' active apostolate, sounds and resounds the simple clear note of Campion's 'Challenge' calling for 'more equitie' towards the oppressed recusant minority. Time and again the word is spoken with direct, unaffected eloquence; and yet, ironically, involved in the term there remains a virtually insoluble ambiguity both of primary definition and of circumstantial application. 'For a long time', F. W. Maitland observes, 'English equity seem[ed] to live from hand to mouth'. He also says that 'no one was prepared to define by legislation what its place should be'. The *OED* article on 'equity' defines it as 'the recourse to general principles of justice (the *naturalis aequitas* of Roman jurists) to correct or supplement the provisions of the law'; but it remains a moot point whether such 'recourse' signifies an attainable legal process or a hypothetical 'notion of an appeal to "higher law"'. There is a particular passage in Southwell's work where the citing of 'equitie' strikes one with the full force of a parable, an exemplary figure bearing witness to a sustainable reality. In the beautiful prose meditation, printed in 1591, *Marie Magdalens Funeral Teares*, the weeping penitent standing before the empty tomb is supposed, for the argument's sake, to consider it an 'impeachment' of her 'right' for Christ thus to 'conuey himselfe away without thy consent'. Since it is a 'rule in the lawe of nature' that a donor cannot 'dispose of his gift without the possessors priuitie', 'thou maiest imagine it a breach of equitie', 'if he hath take[n] a way himself'. In the induction to his long poem 'Saint Peter's Complaint' there are, again, lines which have an emblematic containment:

> If equities even-hand the ballance held,
> Where *Peters* sinnes and ours were made the weightes:

Elsewhere, however, in other sections of *Marie Magdalens Funeral Teares*, in the *Humble Supplication*, also of 1591, in the 'Letter to Sir Robert Cecil', of 6 April 1593, and in the *Short Rules of a Good Life* (1595) Southwell employs the term 'equity' in what we may refer to as an 'appealing' pattern: 'Thogh I were to sue to the greatest tyrant, yet the equitie of my sute is more then halfe a grant'; 'And as the equity of the cause, doth breath courage into the defendors'; '. . . might in equity Challenge all mens penns to warne you of soe perilous Courses'; '. . . and incline you to measure your Censure with reason and Equity'; '. . . so far as with justice and equity they can demand'; 'For though an indifferent arbiter . . . could not in equity disprove my courses'.

According to Hugh Trevor-Roper, the 'great age of the recusants was . . . the age of their greatest dilemma'. The Papal Bull of 1570, *Regnans in Excelsis*, pronouncing sentence of excommunication upon Elizabeth, was sufficiently hasty in conception and 'uncanonical' in its terminology to permit some seeming latitude in the matter of negotiation and compromise. The 'declaration of allegiance' of 1585, submitted by the wealthy and influential Catholic layman Sir Thomas Tresham, was 'an attempt to work out some kind of acceptable division between the spheres of influence which might be covered by a Catholic gentleman's relation to his queen and his relation to his priest'. And for the next twenty years, during which some one hundred and eighty Catholic priests and laymen went to the scaffold and while his own feelings grew increasingly hostile towards the Jesuit mission, Tresham, with exemplary patience or remarkable obtuseness, persisted in his attempts to effect a compromise between conscience and 'acceptability'. Such a compromise was, in fact, unattainable because, as Christopher Morris has argued with cogent irony, the government, in Machiavellian fashion, proceeded 'in accordance with "reason of state", which was somehow different from normal reason'.

When Robert Southwell, having already been tortured several times, was brought for interrogation before the Privy Council, he was complimented on the courtesy of his demeanour and was asked why he had not behaved with equal reasonableness to his tormentor Topcliffe. 'Because,' he answered, 'I have found *by experience* that the man is not open to reason.' This strikes one as being the most crucial of confrontations, the most searching of contexts. The mission-priests faced a situation in which the 'normal reason' of men like Tresham was compromised at every turn by 'reason of state'. Southwell's retort, in an instant, both judges the travesty and redeems the word. Equity was, indeed, forced to live 'from hand to mouth'. Where it chiefly endured, sustained and sustaining, was in printed 'supplication' and 'challenge', in the extempore but nonetheless deeply meditated speeches in court, in enforced 'conferences' and on the scaffold; and, of course, in Southwell's poems. Maitland has said that 'after the brilliant thirteenth century . . . Law was . . . divorced from literature'. Through the nature of a paradox, a glimpse of reunification was made possible, in the age of Elizabeth, by the work of those whose lives were forfeit to a more savage kind of divorce. It has been argued that 'during Shakespeare's lifetime equity was both an important ethical principle and, through the Court of Chancery, an increasingly strong legal force'. *Measure for Measure* has been called a 'masterpiece of comedy on the theme of administering the law with justice and equity'. But for the arraigned priests and laymen there was no Court of Chancery and the only theatre in which they could enact their 'theme' of the weighing of justice with equity was the public arena of controversy, trial and execution. The strength of the creative paradox turns, then, upon the legal helplessness of the petitioners. As Paul Vinogradoff has shown, there was a forensic precedent in Roman Law, going back at least as far as the Bologna law school of the late eleventh century, in which Irnerius defined 'equity' as 'the mere enunciation of a principle of justice'. 'Mere enunciation' in the work of Campion and Southwell had not only to proclaim the inviolability of an ethical principle but also to appear invested in the authority of 'strong legal force', even though its only court of appeal was the appeal of its own eloquence. It is our recognition of this fact which justifies Pierre Janelle's allusion to Southwell's conduct at his trial and execution as being 'a work of art of supreme beauty'.

Our concern is with the style of Robert Southwell, a poet in both verse and prose. Style is not simply the manner in which a writer 'says what he has to say'; it is also the manner of his choosing not to say. There is a distinction to be drawn here between the manner of not-saying and the demeanour of silence. At his trial in 1535 Sir Thomas More had made what R. W. Chambers calls his 'great plea for the liberty of silence'. The Elizabethan missionaries were, in all humility, proud of their silence under torture. The seminary-priest John Ingram, soon to be executed at Newcastle upon Tyne, reported that Topcliffe had called him a 'monster' of 'strange taciturnity'. Robert Cecil told how he had seen Southwell, subjected to 'a new kind of torture', 'remain as dumb as a tree-stump; and it had not been possible to make him utter one word'. This very 'taciturnity' and 'dumbness' are in themselves powerful coadjutors to the eloquence of the polemical and meditative writings. But 'choosing not-to-say', just as much as choice of words, presupposes a 'hinterland' of style, a 'back-country' of what might, for better or worse, have been said. In our necessary exploration of the 'back-country' of Southwell's eloquence we encounter violence and coarse preciosity and disgust, and what he himself significantly referred to as 'wittye crueltye'.

There is a passage in Thomas Nashe's *The Unfortunate Traveller* (1594) describing a public execution:

> The executioner needed no exhortation herevnto, for of his owne nature was he hackster good inough . . . At the first chop with his wood-knife would he fish for a mans heart, and fetch it out as easily as a plum from the bottome of a porredge pot.

A modern scholar has written that 'the pleasure in a job well done palliates some of the unpleasantness of the description'; which seems a cruelly inept comment on Nashe's 'witty cruelty'. At the execution of Fr Edmund Gennings in 1591, 'the martyr crying upon St Gregory his patron to assist him, the hangman astonished said with a loud voice, "God's wounds! His heart is in my hand and yet Gregory is in his mouth"'. There is a world of difference between talk of pleasure palliating unpleasantness and Janelle's suggestion that Southwell made of his own execution 'a work of art of supreme beauty'; it is the difference between collusion and transfiguration. It has already been remarked that the Ignatian exercise of 'seeing in imagination the material place where the object is that we wish to contemplate' must, for Southwell, have involved the 'seeing' of his own almost inevitable martyrdom. In the *Epistle of Comfort* he writes:

> And as a cunninge imbroderer hauinge a peece of torne or fretted veluet for his ground, so contryueth and draweth his worke, that the fretted places being wroughte ouer with curious knottes or flowers, they farr excel in shew the other whole partes of the veluet: So God being to worke vpon the grou[n]de of our bodyes, by you so rente & dismembred, will couer the ruptures, breaches, & wounds, which you haue made, with so vnspeakable glory, that the whole partes which you lefte shalbe highlye beautifyed by them.

An immediate objection would be that Southwell is too much the 'cunning embroiderer', that he 'contriveth and draweth' too 'curiously'; but a more patient consideration would have to concede that the 'ground' upon which he weaves his variations could not be more plainly stated: 'our bodyes . . . rente & dismembred'. Southwell is foresuffering his own agony even as he rises serenely above the fear and the violence: 'Our teares shalbe turned into triumphe, our disgrace into glorye, all our miseryes into perfect felicitye'.

Violence of one kind Southwell not only allows but approves. In his Latin *Spiritual Exercises and Devotions*, which belong to the years of his novitiate, he writes:

> It is a great hindrance to refrain from using violence to oneself, to offer but a feeble resistance to the passions and other obstacles, or to adopt remedies which are almost useless. 'For the kingdom of heaven suffers violence and the violent bear it away.' Moreover experience shows that even the most powerful means are scarcely sufficient for our cure.

This reflection turns upon a scriptural text, Matthew 11:12, which 'had particular significance for Southwell': *Regnum coelorum vim patitur, et violenti rapiunt illud.* This is a text which is recognized by modern commentators to be notoriously 'difficult', 'enigmatic'. Who are 'the men of violence'? The modern 'consensus' agrees that the term must be glossed in a pejorative sense: they are the men of 'hostile intent': Herod, the Pharisees, 'official Israel', the persecutors of both John the Baptist and Christ. This reading would have been so apposite at a time when 'official England' seemed bent on out-Heroding Herod, that one must with some astonishment concede that Southwell, who quotes the text several times, everywhere glosses it in a favourable allegorizing sense. In the *Epistle of Comfort*, where the text is also used as the epigraph on the title-page, he writes 'and though our champions, be of more courage, and our foes more enfeebled, since our redemption, yet doth *the Kingdome of heauen still suffer violence, and the violent beare it awaye*'. There might seem to be a transient ambiguity here; but my feeling is that Southwell resolves it by adding a phrase from 2 Timothy 2: 5, '*and none shall be crowned, but they that haue lawfullye foughte for it*'. The second stanza of the lyric 'At Home in Heaven' also turns upon the Matthean text. Addressing the aspiring soul Southwell writes:

> Thy ghostly beautie offred force to God,
> It cheyn'd him in the lynckes of tender love.

He is in no way at odds here with the reading established by the early Church Fathers or with the scriptural exegesis of his time, whether Catholic or Protestant. Luther had written, quoting this text, 'in my judgment, prayer is indeed a continuous violent action of the spirit as it is lifted up to God'. The so-called 'Evangelical Catholic' Juan de Valdés commented on the text thus: 'if you wish to take the Kingdom of Heaven, do violence to yourself, and so you will fear nothing'. The 'Rhemes' New Testament of 1582 does not annotate the text; but the Jesuit commentator, Cornelius à Lapide, provides a detailed allegorizing gloss. Though only six years younger than Southwell, he cannot truly be considered as being of the same 'generation'. He became Professor of Exegesis at Louvain in 1596, the year after the martyr-poet's death, and his New Testament *Commentaries*, begun in 1616, were not published until 1639. It is not unreasonable to suppose, however, that Lapide's seventeenth-century *Commentaries* drew upon the kind of Biblical exegesis prevalent in Jesuit circles in the closing years of the previous century. Lapide's interpretation states that 'for the Kingdom of Heaven's sake worldly men do violence to themselves by the cultivation of repentance, poverty, continence, mortification'. He further interprets 'violence' as the heroism necessary to endure the sufferings of martyrdom: 'Thus let each believer consider that with his utmost energy he must struggle up to Heaven by means of a ladder hedged about with knives'.

Although Southwell's unquestioning gloss on 'violence' in Matthew 11:12 seems too bland and too partial when read in the light of modern New Testament criticism, in other contexts he clearly associates the word with Herodian savagery and with the disorders of the private will. He writes, in one passage, of 'tiranical persecution . . . most violentlye bent' against Catholics, in another of 'violent tortures', and, in a third, of those who 'haue with violence martyred and oppressed vs'. He depicts the 'violence' of Mary Magdalen's grief as well as her 'violent' love. There is indeed a suggestion that he associates 'violence' with the temptation to do wrong in a good cause: 'if thy Lord might be recouered by violence . . . wouldest thou aduenture a theft to obtaine thy desire'. When, in the brief but most beautiful 'A prayer in temptation', he writes, 'I am urged against my will and violently drawn to think that which from my heart I detest' it is as though we have found the focus upon which these different lines of emphasis converge. One might risk the suggestion that in these radical changes of connotation, Southwell is instinctively probing, more keenly than he would consciously recognize, his own conformity with doctrine and the limits of his own rhetoric; as though in matters theological his poetic vision had a prescient advantage over his theology. But the contrary suggestion is equally tenable: that the range of possibility in 'violence' in no way eludes him; that his method is simply and profoundly eloquent, 'speaking out', 'making clear', the complex hazards of equity:

> [A]s not to feel sorrow in sorrowful chances is to want sense, so not to bear it with moderation is to want understanding; the one brutish, the other effeminate.

Janelle is quite right, however, to suggest that here 'mere reasonableness [is] raised to a divinely spiritual plane'. Southwell's style is equity made palpable; it is also, as I have already implied, an art of 'transfiguration'. The term connotes both 'metamorphosis' and 'elevation'. What, one may ask, is to be transfigured? And one may answer: the violence, the preciosity and disgust, the 'witty cruelty', the 'hinterland' of Elizabethan mannerism and atrocity. In a significant sense, for the devout recusant the medium of 'transfiguration' already existed in the form of the reliquary: 'Of the venerable martyr [Thomas] Bolliquer . . . a little piece of his heart . . . some of his praecordia . . . some papers greased with his fat . . .' Richard Simpson commented in 1857 – and Dom Bede Camm quotes him with approval – 'these relics were not the less venerable on account of the disgusting processes they had gone through; the horror does not attach to them, but to the brutes who presided over the butchery'. This is well said, but the very strength of Simpson's emphasis registers the intensity of the 'horror' and the 'disgust' which are there to be overcome. A manuscript 'Catalogue of Martyrs', written *c.*1594, probably in the hand of Southwell's friend and co-worker John Gerard, records the martyrdom of Fr Thomas Pilchard, 'quartered' in 1587. The 'officers retorninge home, many of them died presently crying out they were poisoned with the smell of his bowells . . . A laye man was executed there some 4 years after . . . whoe beinge asked at his deathe, [what] had moved him to that resolution, etc., he saide, "Nothinge but the smell of a pilcharde"'. This is the kind of thing that Southwell 'lived with' in every sense. In 1584, while still at the English College in Rome, he received, from a friend who had been standing 'under the gibbet', a detailed account of the martyrdom of George Haydock. For Southwell the 'ladder hedged about with knives' could never have been a mere emblem; and, in this, I believe, is one of the sources of his rhetorical strength. To speak of his 'fastidiousness' in the face of such atrocious business might seem to imply a wincing kind of sensitivity. What he attains is rather an eloquent modera- tion, neither 'brutish' nor 'effeminate'. He too could contrive a 'conceit' out of disembowelling but his witty exercise has none of the cruel flippancy of Nashe and differs markedly from the coarse and acrid punning in the 'Catalogue of Martyrs'. In a letter of December 1586 to Fr Agazario, he writes: 'You have "fishes" there [i.e. in Rome] greatly wanted here, which, "when disembowelled, are good for anointing to the eyes and drive the devils away," while, if they live, "they are necessary for useful medicines"'. Southwell's letters from England were sometimes written in a 'veiled style' for reasons of security; but these words are not so much like a 'code' as like a serenely witty form of tact. They resemble what an Elizabethan musician would have called a 'division upon a ground', a variation upon some verses from chapter 6 of the Apocryphal Book of Tobias:

> Then sayd the Angel to him: Take out the entralles of this fishe, and his hart, and gal, and liuer, keepe to thee: for these are necessarie and profitable for medecines . . . If thou put a little peece of his hart vpon coales, the smoke therof driueth out al kinde of diuels, either from man or from woman, so that it cometh no more vnto them.

A recusant martyr's 'entralles' were, of course, 'taken out', his heart was burned, fragments of his body were kept by the faithful as 'necessarie . . . medecines'. Southwell, significantly, adapts his source to suggest the virtues of the living priest as well as those of the 'embowelled' martyr. It is noteworthy too that both the minatory 'exemplum' of Pilchard's bowels and Southwell's 'conceit' of ministration and martyrdom must surely refer to the same scriptural source. The first, however, is pitched at the level of Tudor chapbook or even jest-book; it is, literally, a 'vulgar spectacle'. I have said that Southwell 'lived with' this kind of thing in every sense; and he was also capable of exploiting vulgar spectacle. His *Epistle of Comfort* contains 'a warning to the persecutors' as blatantly *ad hominem* as anything in the 'Catalogue of Martyrs' or Bunyan's *Grace Abounding*:

> Remember the sodayne and horrible deathe of one Yonge an Apostata and Pourswivaunt who pursuing a Catholike at Lambeth fell doune on the sodayne, ere he could laye handes on him that he persecuted and foming at the mouthe presentlye dyed.

Even though, as is well known, Southwell worked mainly among the nobility and gentry, and though Arundel House must have provided an exceptionally cultured spiritual environment, he was nonetheless 'hedged round' with sordid violence; he knew that his own execution would be a 'vulgar spectacle' too. I do not imply that he was 'sceptical' of these popular marvels (indeed, he challenges anyone who might doubt that the Thames stood still on the day of Campion's martyrdom). I do suggest, however, that one of the achievements of his own polemical style is its ability to turn both injustice and 'revenge' in the direction of equity; towards 'the reparation of wrongs rather than the punishment of offences'.

More than one authority has described Southwell's characteristic method as that of 'transformation', and it is a word which matches the terminology of Biblical scholars discussing another of Southwell's texts: Philippians 3:21. 'Transformation' can be read both as an acknowledgement that a significant amount of Southwell's work comprises 'translations, adaptations . . . imitations' and 'parodies' and as a recognition of a process that he himself called 'wonderful alteratio[n]'. Strictly interpreted, this means the radical change in 'mens maners' engendered by the blood of martyrs. Figuratively applied, it could be said to describe a crucial 'turn' which is a feature of his style:

> And this is that which Saint *Paule* sayd: *Reformabit corpus humilitatis nostrae, configuratum corpori claritatis suae:* He shall reforme the body of our humility confygured vnto the bodye of his brightnesse. Whiche phrase of speache argueth, that the more the body for him is humbled in torments, the more shall yt be partaker of hys brightnesse in glorye.

'Reform' is the 1582 'Rhemes' reading of the text which a modern scholar paraphrases as '[He] will refashion our body of lowliness to share the form of his body of glory'. The doctrinal point which is currently stressed – 'that Paul does not think of his eternal blessedness in terms of the separation of the soul from the body' – is not the matter upon which Southwell concentrates. His concern is to affirm both that the 'scattered parts' of the martyrs will be 'reformed', restored, put together again; and that in due proportion as the body is disfigured for His sake on earth so it shall partake of 'transformation' at Christ's Parousia. Southwell's 'phrase of speech' is one that 'argues' for ultimate equity. But again, it is a sense of equity which turns upon a meticulous attention to sequence and context. As he is at pains to make clear, the transfiguration upon Mount Tabor preceded the Passion. It was more a sign of an initiation to extremity of suffering than a reward for suffering endured and transcended. The word 'reason' itself becomes part of the sinew of Southwell's argument here:

> There is no reason, that Christe shoulde shew him selfe more fauorable to vs, that haue bene his enemyes, then to his owne bodye, neyther can we iustlye complayne, if ere we find him, he giue vs a sipp of that bitter chalice, of which for our sakes he was contente to drincke so full a draught. Yea we may be hartelye glad, if after long teares and deepe syghes, we maye in the ende fynde him at all, whether it be in the pouertye of the cribb and maunger, or in the agonyes of his bloodye sweate in the gardeyne, or in the middest of blasphemyes, reproches, and false accusations at the tribunals, or in the tormentes of a shamefull death vpon the Crosse.

Christ 'transfigured in *Mounte Thabor* . . . was also at the same time, heard talkinge *de excessu* of his bitter passion'. Christ 'transfigured', therefore, is not the Christ-in-glory of the Parousia. What Yeats was to call 'Calvary's turbulence' was perhaps better understood by Southwell. His 'turns' are models of, and ways of mastering, the turbulence in 'the air around him' and in his own spirit: 'For Thy sake allow me to be tortured, mutilated, scourged, slain and butchered' he had written in his early *Spiritual Exercises and Devotions*, adding 'I refuse nothing'. These three words are of radical significance: they are the 'wonderful alteration' of a hovering morbidity into a positive oblation.

I have referred earlier to the 'hinterland' of Southwell's style and to the violence and preciosity and disgust which we encounter there. There was also, in Elizabethan as in modern literature, a disgustingly violent preciosity. We are assured that Southwell knew a fair amount of Ovid by heart; and in the *Metamorphoses* the Elizabethan student could find, drawn with 'linger-ing-out sweet skill', the flaying alive of Marsyas by Apollo. In Seneca's *Thyestes*, 'faythfully Englished' in 1560 by Jasper Heywood, the future Jesuit Provincial: 'From bosomes yet aliue out drawne the trembling bowels shake . . .' In John Studley's translation of the same dramatist's *Hippolytus*, the protagonist is trampled and torn to pieces amid the wreckage of his chariot:

33

From bursten Paunch on heapes his blouddy bowells tumble thick . . .

In the Elizabethan 'hinterland', where spectacular 'closet' horror can at any time become the routine hideousness of public spectacle, how can one say where metaphor ends and reality begins?

This weak rhetorical question is out of keeping, however, with Southwell's power to distinguish and affirm. At the heart of his own eloquent style stands a patristic text which, in the *Epistle of Comfort*, he attributes to St Cyprian confronting his persecutors:

> Whye doest thou turne thee to the fraylty of our bodyes? Why stryuest thou with the weaknesse of our fleshe? Encounter with the force of our minde; impugne the stoutnesse of our reasonable portion; disproue our faythe; ouercome vs by disputation if thou canst, ouercome vs by reason.

It is here that we encounter the paradigm for that 'absolute reasonableness' to which my argument alludes. The 'force of our minde' is a key-term for Southwell's form of argument as Donne's 'masculine perswasive force' is a key-term for his; but the differences of implication between the two phrases are greater than any apparent similarities. For Southwell, 'force of . . . minde' is manifested in the power to remain unseduced and unterrified, whereas Donne's words relish their own seductive strength. Dame Helen Gardner has fairly remarked that Donne forbids us to 'make any simple equation between the truth of the imagination and the truth of experience'. If that is so we may regard Southwell as Donne's antithesis, for his constant practice is to show 'how well verse and vertue sute together' and the 'simple equation' which Donne precludes is the 'equity' to which Southwell appeals in phrases which are like emblems of his faithful reason: 'Whose measure best with measured wordes doth sit'; 'Where vertues starres God sunne of justice is'; 'And though ech one as much as all receive, / Not one too much, nor all too little have'.

The correlative of equity is sacrifice and Southwell sacrifices a great deal, even the poet's delight in self-sustaining, self-supporting wit. Helen C. White has written of his predilection for 'Baroque . . . transformation' but, as I see it, the matter turns more upon his sensitivity to a secular domain of 'baroque' inequity, of 'uneven accompt', which he seeks, in his poetry as in his prose, to 'reform' to a 'just measure': 'the affections ordinate, and measurable, all the passions gouerned by reason, and settled in a perfecte calme' and 'in the syghte of God'. It is entirely characteristic of Southwell's art, however, that, 'ordinate and measurable' though it is, it brings us face to face with violent contradictions. We are bound to assent both to its mediocrity and its monotony, for 'mediocrity' is essentially nothing more nor less than 'measured conduct or behaviour' and 'monotony' is 'sameness of tone or pitch'. If, however, we take this latter term to mean 'wearisome sameness of effect, tedious recurrence of the same objects, actions, etc.', we give our assent to Southwell's indictment of our carnal world. In his eyes it is a world vacuously full of 'loathed pleasures', 'disordred order', 'pleasing horror', 'balefull blisse', 'Cruell Comforts'. The existence of the carnal sinners is an oxymoronic treadmill; and their only means of redemption is by way of the divine paradox. As J. R. Roberts reminds us, Saint Ignatius was particularly 'awestruck' 'by the fact that the Creator had become man' and Southwell is in this, as in all things else, Ignatian. God disadvantaged himself for man's advantage and the priest-poet is concerned to stress both sides of the redemptive equation:

> This little Babe so few dayes olde,
> Is come to ryfle sathans folde;
> All hell doth at his presence quake,
> Though he himselfe for cold doe shake:
> For in this weake unarmed wise,
> The gates of hell he will surprise.

Jeffrey Wainwright draws attention to the 'naïve though winning concession' which he finds in one of Southwell's turns of phrase ('Passions I allow, and loues I approue . . .'). 'Winning' is exquisitely apt. 'Naïve' calls for more care; but is justifiable, I believe. Wainwright directs us to the crux of Southwell's circumstance and achievement. Southwell is 'naïve', if 'witty cruelty' is the world's alternative; but if he is so, it is by choice; and the choice is doubly purposeful. His 'naïvety' is to some extent penitential, submitting the upstart creative 'will' to 'bonds' of humility; and it is, to a further extent, evangelistic. His Nativity Poems, wholly in keeping with Ignatian precept, as Roberts has shown, weigh the harshness and the tenderness of the scene at Bethlehem in order to win men to repentance and love.

One cannot choose 'naïvety', however, if one has no sophistication from which to turn. Southwell, as Janelle and Thurston have amply demonstrated, was a highly sophisticated master both of classical rhetoric and of the modern Euphuistic style. In 'New heaven, new warre' such words as 'ryfle' and 'surprise' are placed with a beautiful tact; and though the rhyme 'quake/shake' may sound naïve we should not assume that naïvety is the cause. That 'complexity of association' which Helen C. White detects elsewhere in Southwell's writings is here the 'ground' upon which he 'works', instead of 'curious knots and flowers', the lilt of a child's catechism. The paradox of the naked new-born babe, the shivering child who shivers the gates of hell, is at the heart of his vision. It is ultimately a vision of great serenity, but, as I have tried to show, that serenity is achieved in the full awareness of the realities of spiritual and legal violence.

There is a sad irony in the fact that Antonin Artaud, godfather of the 'theatre of cruelty', translated Southwell's 'The Burning Babe' ['Le Bébé de Feu'] during his incarceration in the asylum at Rodez. Southwell would not grudge him that grace; but it would be a matter for regret if the violent preciosities of extremism were to set their seal of approval on Southwell's profoundly different understanding of the condition of both the tormented and the ecstatic soul. 'Christe was . . . heard talkinge *de excessu* of his bitter passion'. 'Excessus' signifies 'ecstasy'. The word was so used by Saint Bernard, the author closest perhaps to Southwell's heart, whose works he cited in *An Epistle of Comfort*; which were also his 'solace' during his imprisonment in the Tower. We may seem here to have returned to J. R. Roberts's suggestion with which we began, that Southwell, in 'The Burning Babe', appears to have 'lost himself in ecstatic delight'. The circle is not quite closed, however. The 'pretty Babe all burning bright', the Christ Child, is 'scorched with excessive heate' which emanates from his own 'faultlesse breast'. There is certainly more 'complexity of association' here than either Janelle or the Clarendon text editors allow. He dismisses it as 'the most hackneyed of all conceits'; they properly point out that it 'parodies . . . Petrarchan tradition'; but something further needs to be said. One would read 'excessive' heat simply in the *OED*'s 'neutral' sense, i.e. 'exceeding what is usual' not 'exceeding what is right', if one were not so aware of the striking parallels, detected by Martz, between this poem and such contemporary Jesuit devotional exercises as Puente's *Meditations* . . . The Christ Child may indeed be talking 'de excessu' from amid a fiery ecstasy of sacrificial love; but I am unable to share Roberts's view that Southwell himself is 'lost . . . in ecstatic delight'. There is a note of deliberate naïvety in the poem which substantiates Martz's suggestion that it should be read as a 'variation . . . on the medieval nativity-ballad, done after the Jesuit manner'. This fact would in no way preclude a sensitive apprehension of the nature of ecstatic experience. St Ignatius himself was much influenced by mystical tradition. But deeply versed though Southwell may have been in the methodology of 'excessus', he was alert to the dangerous implications of excessive behaviour. 'Excesse of minde' is the 'Rhemes' translation of the term in Acts 10: 10, which the Authorized Version renders as 'trance'. Even so, Southwell, in his *Short Rules of a Good Life*, wrote that 'excess in the voice and immoderate loudness are always certain signs of passion and therefore ought to be used but upon some extraordinary necessity'. For Southwell to have 'lost himself' merely in a poem would have required more self-centredness than he was capable of.

This keenly witty man who was so properly sceptical of 'fancie' and 'selfe delight' employed all the resources of his wit to moderate between grace and peril in this most dangerous area of the religious life. 'Let vs but consider, the last tragicall pageant of [Christ's] Passion, wherein he wone vs, and lost him selfe. And marke the excessiue loue shewed therin . . .' he wrote in *An Epistle of Comfort*. 'Excess', 'excessive', 'de excessu', 'Passion', 'passion', 'violence', 'equitie': these are all cruces, the little crosses upon which the passion of his reasonableness is enacted for us. We have it on the authority of Fr Christopher Devlin that 'the Jesuit discipline, in the design of St Ignatius, sets up an interior tension which can only be resolved by crucifixion. At the heart of it there is an element of supernatural wildness . . .' In the words 'discipline' and 'wildness' we confront that paradox which Southwell perfectly understood, as I believe he also understood the fulcrum of Ignatian 'design'. I would further suggest that the radical pun perceivable in 'ecstasy', in being 'beside oneself', either with a frenzy of egoistic inclinations or with a disciplined indifference to them, would not be lost on him. When he was brought out to endure 'the torments of a shameful death' Southwell could speak, with perfect calm and tact, in the idiom of his own *Epistle of Comfort*: 'I am come hither to play out the last act of this poor life'. Even at that moment he could retain his grasp on 'complexity' and yet speak with absolute simplicity. And it was such complex simplicity, I would finally claim, that enabled this man of discipline to concede, in *Marie Magdalens Funeral Teares*, the 'wonderful alteration' of 'wildness' itself: 'Loue is not ruled with reason, but with loue'.

37

THREE

'The World's Proportion':
Jonson's Dramatic Poetry in *Sejanus* and *Catiline*

The worlds proportion disfigur'd is,
That those two legges whereon it doth relie,
Reward and punishment are bent awrie.

Jonson's two Roman tragedies have seemed to many readers to be harsh, intractable and unrewarding. From time to time accusations of soulless pedantry have been made against their author. According to Sir John Sandys:

> Jonson is constantly hampered by his learning . . . In his *Catiline* and his *Sejanus* we find elaborate notes from Sallust and from Tacitus respectively, to prove he had good authority for every detail in his drama, and to make it perfectly plain that 'Fancy had no part in his work'. While Jonson thus preserves the outer garb of the Roman world and allows the soul to escape him, Shakespeare makes his men true Romans . . .

Both Coleridge and Swinburne, however, have intelligent notes on the relationship of these plays to the Jacobean political dilemma. They are at least prepared to accept that Jonson's choice, and treatment, of his themes may have been governed by something other than scholarly perverseness.

In both *Sejanus* and *Catiline*, Jonson manages to blend a forthright dogmatism with an astute trimming. This makes him, in a way, an epitome of the disturbed decades preceding the outbreak of the Civil War. This was the time of Selden, the troubled democrat; and of Falkland, the troubled Royalist; and Jonson knew them both. And, at a period when the tensions between Court and City were becoming increasingly felt, Jonson abandoned the public theatres for ten years to become a writer and producer of masques for royalty and aristocracy. A despiser of the crowd, he yet said 'Words are the Peoples' and captured in his comedies the idioms and inflexions of common speech. And modern critics have, of course, sufficiently appraised Jonson's 'accurate eye for . . . oddities', his 'quick ear for the turns of ordinary speech'. Nevertheless, this very admiration produces its own kind of danger. To suggest that art, to be significant, must grow from vernacular roots is so much a contemporary necessity that it becomes, almost, a required cliché. And to say that, in Jonson,

> the tricks of shysters and crooks, mountebanks, lawyers, news-vendors, and monopoly-hunters are transferred to the stage with all the relish of one who sees for himself what is under his nose

is to make his satire seem more indulgent than it is. 'Relish' and 'under his nose' suggest a slightly myopic gourmet.

The recent and thorough evaluation that the comedies have received from various hands appears to leave a residual implication: that compared to the tough contact with life in plays like *The Alchemist* or *Bartholomew Fair*, the world of the Roman tragedies is an abstraction, a retreat into formulas and commonplaces. The cadences and imagery of these plays are certainly at a remove from those of Busy or Dol Common; but it could be argued that the grand commonplaces – even those that refer back to Latin originals – are as much part of the seventeenth-century climate as are the idioms of the great comic characters. It seems to be a modern fallacy that 'living speech' can be heard only in intimate situations; in fact the clichés and equivocations of propaganda or of 'public relations' are also part of the living speech of a society. Character writers of the seventeenth century used Roman allusions, as modern political cartoonists employ a kind of visual shorthand, to elicit a required stock response. It could be argued that the high-pitched invective that runs through *Sejanus* and *Catiline* is superb 'cartoon' language, with all the serious overstatement and polemic relevance of that art:

> CICERO. What domesticke note
> Of priuate filthinesse, but is burnt in
> Into thy life? What close, and secret shame,
> But is growne one, with thy knowne infamy?

The vituperative 'filthinesse', 'shame', 'infamy' could, perhaps, be taken too easily as a mere stage-anger, as an abstraction from the tensions, the 'humanity', of real experience. The stresses of Jacobean and Caroline society, culminating in the Civil War, were real enough; yet men declared their living angers, their immediate and pressing problems, in a language comparable to that of *Catiline*. The *Declaration of the County of Dorset, June 1648* was framed in the following terms:

We demand . . .

(iv)   That our Liberties (purchase of our ancestors' blood) may be re-deemed from all former infringements, and preserved henceforth inviolable; and that our ancient liberties may not lie at the mercy of those that have none, nor enlarged and repealed by votes and revotes of those that have taken too much liberty to destroy the Subjects . . .

(vii)   That we may no longer subjugate our necks to the boundless lusts and unlimited power of beggarly and broken Committees, consisting generally of the tail of the gentry, men of ruinous fortunes and despicable estates, whose insatiate desires prompt them to continual projects of pilling and stripping us . . .

The 'boundless lusts' and 'ruinous fortunes' anathematized in the *Declaration* may be set against Cicero's denunciation of Catiline and his fellows, their 'filthinesse', their 'shipwrack'd mindes and fortunes'.

Jonson shares with the authors of the *Declaration* a dual attitude towards the name and nature of liberty. The 1648 petition puts liberty in two poses: the first respectable; the second disreputable. 'Liberty' in the first case is plainly involved with property and heredity ('purchase of our ancestors' blood'). Opposed to this is a second, destructive 'liberty', synonymous with licence ('boundless lusts', 'insatiate desires'). In Jonson's moral world, also, a stress and counter-stress is evoked from the conflicting connotations of words such as 'liberty' or 'freedom'. In the Roman tragedies 'liberty' seems frequently equated with irresponsible power:

> CAT[ILINE].   Wake, wake braue friends,
> And meet the libertie you oft haue wish'd for.
> Behold, renowne, riches, and glory court you . . .
> And, being *Consul*, I not doubt t'effect,
> All that you wish, if trust not flatter me,
> And you'd not rather still be slaues, then free.
> CET[HEGUS]. Free, free. LON[GINUS]. 'Tis freedom.
>   CVR[IVS]. Freedom we all stand for.

As parallel to this grotesque shout there is, in *Sejanus*, the 'Senates flatterie' of the new favourite Macro; their weak cries of 'Liberty, liberty, liberty' as he takes Sejanus' place. The irony here is that the Senators cry out to celebrate the return of that liberty which is a concomitant of property and stability; but we know that 'liberty', under Macro, will still be only 'licence' – a renewal of 'pilling and stripping'.

The civic alternatives to such dissolute, destroying forces are 'industrie', 'vigilance' and a proper reverence for the gods. The Second Chorus in *Catiline* speaks of the need 'to make a free, and worthy choice'; and the good soldier, Petreius, instructs his army in the fundamentals for which it is fighting:

> PETREIVS.   The quarrell is not, now, of fame, of tribute,
> . . . but for your owne republique,
> For the rais'd temples of th' immortall gods,
> For all your fortunes, altars, and your fires,
> For the deare soules of your lou'd wiues, and children,
> Your parents tombes, your rites, lawes, libertie,
> And, briefly, for the safety of the world.

It is clear from this that the word 'liberty' regains, in the oratory of Petreius, its respectable associations, being securely anchored next to 'lawes' and preceded by the pietist sequence: 'gods', 'altars', 'wiues', 'children'. There is danger in the handling of such deliberately emotive catalogues. At worst, such rhetoric can become a confusion of apparent humility and actual hectoring. William Wordsworth, a late democrat beginning the drift back to conservative orthodoxy, employs a similar imagery in his sonnets of 1802:

> altar, sword and pen,
> Fireside, the heroic wealth of hall and bower . . .

Wordsworth's invocation is more raucous than Jonson's, being somewhat burdened by the self-imposed Romantic role of poet-as-prophet. Jonson, exploiting the dramatic medium, is able to distance his orthodox eloquence by projecting it through the mouth of a persona.

Elsewhere, too, the essentially conservative rhetoric is handled with persuasive charm, as in the *Catiline* dedication to the Earl of Pembroke:

> MY LORD,   In so thick, and darke an ignorance, as now almost couers the age, J craue leaue to stand neare your light: and, by that, to bee read. Posteritie may pay your benefit the honor, & thanks: when it shall know, that you dare, in these Iig-giuen times, to countenance a legitimate Poeme. J must call it so, against all noise of opinion: from whose crude, and ayrie reports, J appeale, to that great and singular faculty of iudgement in your Lordship, able to vindicate truth from error. It is the first (of this race) that euer J dedicated to any person, and had J not thought it the best, it should haue beene taught a lesse ambition. Now, it approcheth your censure cheerefully, and with the same assurance, that innocency would appeare before a magistrate.

This is a well-contrived amalgam of religious, political and literary connotations. The key-words embody dual associations of order, authority, and harmony in more than one field. 'To vindicate truth from error' sounds like Hooker; and 'iudgement' and 'innocency' have theological overtones, even though Jonson's prime application is in secular self-vindication. 'Innocency' thus serves to fuse the ideas of divine order and human law. 'Censure' is the province of priest and magistrate and satirist; it is also the privilege of master-patron to servant-poet as Jonson suggests. The dedication is not only an extremely bland and suasive piece of self-defence; it is also a testament to conservative legitimist order; within the bounds of such prose, the tenure of kings and magistrates is unquestionably secure. The very 'assurance', the intuitive rightness of Jonson's usage, stems from a familiar rhetoric, a vision of order typified by Sir Thomas Wilson:

> I thinke meete to speake of framing, and placing an Oration in order, that the matter beeing aptly setled and couched together: might better please the hearers, & with more ease be learned of al men. And the rather I am earnest in this behalf, because I knowe that al things stande by order, and without order nothing can be.

Again, in Jonson's own 'Epistle to Selden', the statesman is praised for his 'manly elocution', a phrase in which a way of speaking and a way of living become one and the same.

Such evocative vindications of wholeness exist, of course, in a social and dramatic context where the unity of ethic and action is a rarity. For the world of Jonson's Roman tragedies is a world 'bent awrie', distorted chiefly by the perverted lust for private gain at the expense of public good. Against all the evidence of tawdry chaos Jonson poses the orthodoxy of Cicero and Cato: rather as Pope, in the *Epistles to Several Persons*, celebrates the decency of agrarian Order in the face of the Whig financiers; and as Yeats sets the 'great melody' of Burke against all manifestations of imperial Whiggery.

In both Roman plays Jonson's vision of moral and civic disorder – though embodied in several forms – is most immediately presented as a reversal of roles in man and woman; as an abdication of power and choice by the former, a usurpation of private and public control by the latter. In *Catiline* the aborted natures of the protagonists are stressed. Sempronia has a very 'masculine' wit; she is 'a great stateswoman', a 'shee-Critick' who can also 'play . . . the orator'. And, in her conversation with Fulvia she acts and sounds like Brutus in his tent:

> SEM[PRONIA].   I ha' beene writing all this night (and am
> So very weary) vnto all the *tribes*,

42

> And *centuries*, for their voyces, to helpe CATILINE
> In his election. We shall make him *Consul*,
> I hope, amongst vs . . . FVL[VIA].   Who stands beside?
> (Giue me some wine, and poulder for my teeth.
> SEM.   Here's a good pearle in troth! FVL. A pretty one.
> SEM.   A very orient one!) There are competitors . . .

The derangement is here stressed by the abrupt parenthesis of womanish trivia, chatter about pearls and dentifrice. Sempronia's strength is an errant thing, a matter of coarse presumption and gesturing.

As moral and dramatic corollary to this 'roaring girl' is the sharp depiction, throughout the play, of lasciviousness and hysteria in the men of Catiline's party. There is the incident of the homosexual attempt on one of Catiline's pages, an episode which so disturbed Coleridge that he suggested emending it out of existence; he called it 'an outrage to probability'. This is precisely its point! The nineteenth century preferred a half-remorseful majesty in its great apostates, a power which Jonson is not prepared to depict in the Catiline conspiracy. Catiline is not Captain Ahab. His rebellion *is* an outrage to probability where 'probability' is more or less synonymous with 'Nature' and the 'moral law'. What Jonson does give is a dramatic relevance to outrage and abnormality; and where the actions of the protagonists seem most to elude the patterns of reason, that is the point of Jonson's attack. These opponents of Cicero are 'men turn'd *furies*', whose 'wishings tast of woman'; into whose mouths Jonson puts a deliberately grotesque hyperbole:

> CET[HEGUS].   Enquire not. He shall die.
> Shall, was too slowly said. He'is dying. That
> Is, yet, too slow. He'is dead.

In this projection of a world of spiritual and physical disorder the Roman tragedies are close to certain of the comedies. In *Epicoene* (1609), varied strata of imagery and action are explored, to discover intricate evidence of ethical and sexual abnormality; of dangerous (even if trivial) perversion in the private and public functioning of human natures. Sejanus, too, was once

> the noted *pathick* of the time . . .
> And, now, the second face of the whole world.

And here the theme of sexual inversion is absorbed into that of broken status, debauched hierarchy. It was by trading his body that Sejanus rose from serving-boy to be potential master of Rome.

43

In *Catiline*, the play of action and imagery, though less intricate than the machinations of *Epicoene*, has its own manner of involvement. Apart from the logical balancing of illogical sexual conditions in men and women, there is a recurring pattern of significant epithets. The conspirators are described (even self-described!) as 'needy' and 'desperate', 'wild', 'lost'. But they are also shown to be possessed by 'sleep' and 'sloth'. At moments of crisis they shout like excited women, or make passes at boys. To be both 'wild' and 'sleepy', 'desperate' yet 'slothful', is gross indecorum, even in rebellion's own terms. It is a double inconsistency, a double irony.

In fact Jonson's method, in both the Roman plays, might be termed an anatomy of self-abuse. It is clear that the *Catiline* conspirators are destroyed as much by their own grotesque self-contradictions as by the bourgeois virtue of Cicero. And in both plays a dangerous apathy or rashness on the part of the body-politic is a contributor to the tragedy. Rome, 'bothe her owne spoiler, and owne prey', is always as much sinning as sinned against. Jonson has Sejanus say:

> All *Rome* hath beene my slaue;
> The *Senate* sate an idle looker on,
> And witnesse of my power; when I haue blush'd,
> More, to command, then it to suffer . . .

At first sight this might appear an inappropriate hyperbole, since Sejanus is presented throughout as a blushless villain. It is certainly no sign of regeneration on the part of the evil-doer; but rather a late out-cropping of the morality-technique. Sejanus, though villain, is allowed an objective condemnation of the ills of the State, is even, for a moment, permitted a Ciceronian touch. Compare:

> CIC[ERO]. O *Rome*, in what a sicknesse art thou fall'n!
> How dangerous, and deadly! when thy head
> Is drown'd in sleepe, and all thy body feu'ry!
> No noise, no pulling, no vexation wakes thee,
> Thy *lethargie* is such.

In some respects Jonson has Sejanus fill a dual role. He blends some of the characteristics of 'Scourge' – bringing torment to a corrupt world – with all the attributes of the *arriviste*, the supposedly amoral disrupter of society's settled decency. As *Sejanus* progresses, Jonson's moral scathing is directed more and more against the blushless dereliction of Senate and people; their abandonment of moral choice. Rome displays the vice of apathy rather than the virtue of patience. This deliberate duality is again stressed at the end.

44

Although it would be quite inappropriate to speak of Jonson's 'sympathy' for Sejanus, nevertheless the drive of the satire is here against society at large rather than against the solitary villain. There is dramatic pity for Sejanus and his children at this point; not because, as human being, Sejanus deserves it, but because his immediate role as mob-victim makes such treatment necessary.

*   *   *

Jonson's dramatic rhetoric, in these two Roman tragedies, is so constructed as to work on two levels, yet to a single end, a comprehensive moral effect. The hyperbole of the protagonists is often so excessive as to be a parable of the spiritual and physical debauchery. At the same time, varied sequences of key-words signal the ethical truth of the action, like spots of bright marker-dye in a greasy flood. In Catiline's wooing of Aurelia, the satiric method is one of profound parody:

> CAT[ILINE].   Wherefore frownes my sweet?
> Haue I too long beene absent from these lips,
> This cheeke, these eyes? What is my trespasse? speake.
> AVR[ELIA].   It seemes, you know, that can accuse your selfe.
> CAT.   I will redeeme it . . .
> AVR.   You court me, now. CAT. As I would alwayes, Loue,
> By this *ambrosiack* kisse, and this of *nectar* . . .

Here is an epicene savouring, a ludicrous distinction between flavours in the kisses, each one mincingly appropriated by the demonstrative. It is the absurdity of over-meticulousness. It is the Spenserian-Petrarchan sexual reverence employed in a context which makes its very euphemism obscene. Marvell adapts the method in 'To his Coy Mistress':

> Two hundred to adore each Breast

where the point is to satirize the idea of virginity as good bargaining-material. In Marvell, as in Jonson, the perspective requires the utterance of deliberate cliché, but cliché rinsed and restored to function as responsible speech. As the lover in *Amoretti* asks for 'pardon' and 'grace' from the beloved and fears lest he 'offend', so Catiline and his paramour toy with a sacramental idiom ('trespasse', 'accuse', 'redeeme'). The suggestive rhetoric is finely used. Catiline is made to utter an effective *mélange* of portentousness and offhand cynicism. We know that he takes himself too seriously and humanity too lightly (his Petrarchan 'fidelity' is pasted nicely upon a cynical confession of wife-removal). The total irony of this scene, of course, depends on the fact that this tiny Petrarchan mock-trial is the nearest Catiline ever gets to an understanding of trespass, or of redemption. He and Aurelia utter, unknowingly, a Punch-and-Judy burlesque of the play's great meanings. It is significant that this evocative scene should be an act of courtship. It is through such sexual attitudinizing and mutual titillation, through such an exposé of the mechanics of allurement, that Jonson presents his broader commentary on the corrupt practices of self-seeking and preferment. The forwardness of the female, the decorative word-spinning of the male, are far from being dramatically inconsistent. And though the gaudy Petrarchan is elsewhere seen as a cold destroyer it is, in fact, the pervading odour of blood that gives a poignant ruthlessness to Catiline's dalliance, his calling Aurelia 'sweet'.

As in Marvell's 'To his Coy Mistress', or Pope's *An Epistle to Dr. Arbuthnot*, Jonson's language is frequently literary in the best sense of the term. That is, its method requires that certain words and phrases, by constant repetition in popular literary modes, shall have been reduced to easy, unquestioned connotations. These connotations are then disturbingly scrutinized. Pope's:

> Oblig'd by hunger and Request of friends

requires for its effect the common formula of gentlemanly apologia on the part of coy amateurs bringing out verse. It is 'hunger' that blasts the cliché into a new perspective. Marvell's wit feeds off Suckling's 'Against Fruition' as much as it does off Marlowe's 'Passionate Shepherd'. Jonson needs the Petrarchan mask even while, in the Catiline-Aurelia scene, he tears it into paper streamers. And, in the following soliloquy from *Sejanus*, Jonson is able to blend the authoritarian tones of Hall-like satire with suggestions of tongue-in-cheek duplicity on the part of the speaker. Sejanus has just concluded his scene of apparently forthright discussion with Tiberius and is alone:

> SEIANVS. Sleepe,
> Voluptuous CAESAR, and securitie
> Seize on thy stupide powers, and leaue them dead
> To publique cares, awake but to thy lusts.

This holds good on two levels. Tiberius *is* voluptuous and does neglect public affairs (though not nearly so much as Sejanus imagines). So far, Sejanus' words may be taken as choric, as in a morality or a satire, setting the scene for the imminent fall of princes. The difference is that Sejanus is gleeful rather than sorrowing or indignant, and that the 'prince' whose fall is imminent is not Tiberius. Tiberius is, in fact, far from stupefied, and succeeds in shepherding Sejanus to the slaughter. Hence, Sejanus' choric comment (still true in its essential commentary on an evil prince) becomes also a statement, by implication, of the fatal blindness of hubris, of Sejanus' own recklessness and greed. In the light of what the audience knows but Sejanus does not, the imperatives of the confident puppet-master ('Sleepe, voluptuous Caesar') slide into the subjunctives of a wishful-thinker ('and [may] securitie seize on thy stupide powers . . .').

From a study of such examples one would conclude that Jonson is able to employ ambiguity – of word, phrase or situation – to give what is ultimately a quite unambivalent expression to moral preference or decision. There is, however, a secondary form of ambiguity at work in both plays – more subjective, more cunning even; and with roots in the Janus-like situation of the professional moralist in Jacobean England.

In *Sejanus*, for instance, crucial political implications are guided into a kind of dramatic cul-de-sac; as in the scene where Sabinus is tempted into self-betrayal by Sejanus' hirelings:

> SAB[INUS].   No ill should force the subiect vndertake
> Against the soueraigne, more then hell should make
> The gods doe wrong. A good man should, and must
> Sit rather downe with losse, then rise vniust.
> Though, when the *Romanes* first did yeeld themselues
> To one mans power, they did not meane their liues,
> Their fortunes, and their liberties, should be
> His absolute spoile, as purchas'd by the sword.
> LAT[IARIS].   Why we are worse, if to be slaues, and bond
> To CAESARS slaue, be such, the proud SEIANVS!
> He that is all, do's all, giues CAESAR leaue
> To hide his vlcerous, and anointed face,
> With his bald crowne at *Rhodes* . . .

The opening of Sabinus' argument is orthodox, reflecting Tudor statecraft as expressed in the Tenth Homily (1547), and anticipating the Canons of 1606 and the Laudian Canons of 1640. James I's considered opinion that 'it was unfit for a subject to speak disrespectfully of any "anointed king", "though at hostility with us"', gives the orthodox viewpoint succinctly enough. Sabinus' acceptance of the official dogmas comes out even in the pithiness of the trotting couplets with which his speech opens. The thought clicks neatly into place together with the rhyme. Then, as a kind of conditional afterthought, Sabinus suggests that very scrutiny of the tenure of kings which he appears to reject at the beginning. The Tudor apologist suddenly begins to speak like a moderate post-Restoration Whig, one to whom: 'government exists, not for the governors, but for the benefit of the governed, and its legitimacy is to be judged accordingly'. Compare, too, the tone of the opening three lines ('hell', 'the gods') with the ironic use of 'absolute' to qualify, not 'kingship', but 'spoile'. Dramatically speaking, this argument (absolutist dogma and Whig qualification) is delivered with the appeal of a Lost Cause. Sabinus is innocently speaking his own doom; his 'fault' is predetermined; his stage audience is wholly composed of *agents provocateurs* and concealed spies, so that his appeals to civic rectitude are sealed off from directly influencing the outcome of the tragedy. It is the theatre audience or the reader ('invisible' witnesses) who are exposed to the full vision of the noble conservative humiliated and betrayed by exponents of corrupt power, of cynical *realpolitik*. And it is arguable that the celebrated line:

> To hide his vlcerous, and anointed face

gains much of its power through a suspension (rather than a union) of opposites. It is at once a well-turned translation of Tacitus' Latin, and an ambiguous metaphor juxtaposing two hostile connotations. These are: 'anointed' in the sense of 'smeared with (ointment or cosmetic)' and 'anointed', meaning 'marked with sacramental authority' (compare 'The Lord's anointed'). Hence, the phrase pivots Jonson's two most powerful social attitudes – disgust at corruption and reverence for consecrated power – thus: 'Tiberius is corrupt; nevertheless, he is the anointed of the gods'. 'Face' continues the deliberate ambivalence; it is both physiognomy and 'cheek' (one is reminded of the character called 'Face' in *The Alchemist*); 'bald crown' signifies both 'hairless head' (a private impotence) and 'stripped authority' (a public impotence). As these words are spoken by Latiaris, the hireling, only in order to trap Sabinus they cannot, therefore, be taken as active radicalism. The image is securely bedded down in a web of tangential implication.

This scene in particular makes nonsense of the view of Jonson's 'originally non-dramatic' nature. He so deploys his conservative philosophy as to give it the attraction of Resistance idealism. He commits himself less deeply than, say, Marvell, who, in 'An Horatian Ode Upon Cromwel's Return from Ireland', speaks – in his own voice, and without dramatic distancing – of Charles's 'helpless Right' and of the 'bloody hands' of the 'armed Bands' round the scaffold. And Jonson's dramatic cunning is also in significant contrast to the procedure of 'Variety', a poem by John Donne or one of his contemporaries:

> The golden laws of nature are repeald,
> Which our first Fathers in such reverence held;
> Our liberty's revers'd, our Charter's gone,
> And we made servants to opinion,
> A monster in no certain shape attir'd,
> And whose originall is much desir'd,
> Formlesse at first, but growing on it fashions,
> And doth prescribe manners and laws to nations.
> Here love receiv'd immedicable harmes,
> And was dispoiled of his daring armes . . .
> Onely some few strong in themselves and free
> Retain the seeds of antient liberty,
> Following that part of Love although deprest,
> And make a throne for him within their brest,
> In spight of modern censures him avowing
> Their Soveraigne, all service him allowing.

The key-words here are like those in Jonson's Roman plays: 'Liberty', 'free'. But it is clear that their significance is far from the 'Ciceronian' ideal. This poet's 'liberty' is not the state of civic wisdom based on 'industry and vigilance', nor is it a concomitant of inherited property. 'Strong in themselves and free' smells more of the dangerous anarchism that, in *Catiline*, yearns to use the 'free sword'. The implications of 'Our liberty's revers'd' and 'Retain the seeds of antient liberty' sing out much more sharply and defiantly in 'Variety' than do the equivalent but well-buffered iconoclasms in Jonson. And, at the crux of his argument, the poet, in offering fealty to a Soveraigne clearly not 'anointed of the Lord', commits a kind of witty high treason. In his own voice he pays homage to 'libertine' nature in terms that Shakespeare preferred to put into the mouth of a branded villain, the bastard Edmund in *Lear*. And, in Jonson, this kind of 'freedom' belongs to degenerate conspirators, rather than to Cecil-like 'new men' such as Cicero. Both Jonson and Shakespeare, though prepared to give an airing to subversive statements in their work, tend to contract out of direct commitment; whereas the author of 'Variety' pursues, with a cynical faithfulness, the conclusions made inevitable by his accepted premises. It could be argued that such distinctions are the product of the time; distinctions between 'revolutionary' ideas circulating in manuscript verses among a small, cultured élite and ideas handled in the drama, a medium struggling against official censorship, and always in the blast of public scrutiny and comment. The ambivalence is certainly there. It was noted by Coleridge, when he said of *Sejanus*:

This *anachronic* mixture of the Roman Republican, to whom Tiberius must have appeared as much a Tyrant as Sejanus, with the *James-and-Charles-the-1st* Zeal for legitimacy of Descent, is amusing.

*   *   *

In *Catiline*, Jonson's treatment of the 'Ciceronian' virtues, and his attitude to the Cicero-Catiline struggle provide a working example of the dramatist's capacity for suspended judgment. Sempronia describes Cicero as:

> A meere vpstart,
> That has no pedigree, no house, no coate,
> No ensignes of a family? FVL. He'has vertue.
> SEM.   Hang vertue, where there is no bloud: 'tis vice,
> And, in him, sawcinesse. Why should he presume
> To be more learned, or more eloquent,
> Then the nobilitie? or boast any qualitie
> Worthy a noble man, himselfe not noble?
> FUL.   'Twas vertue onely, at first, made all men noble.
> SEM.   I yeeld you, it might, at first, in *Romes* poore age;
> When both her Kings, and *Consuls* held the plough,
> Or garden'd well: But, now, we ha' no need,
> To digge, or loose our sweat for't.

Jonson so places the virtue of old Rome in the mouth of a 'modern' degenerate that he gets away with a good deal. To dig or to lose one's sweat honestly at the plough becomes the sweet untainted antithesis to the 'sloth', 'fatness', 'lethargy' of the present. Any scepticism we might have had regarding the tireless virtue of simple poverty is ruled out of court together with Sempronia's gibes. All that she sees as laughable or contemptible we are to receive as serious and worthy. 'Romes poore age' is unquestionably Rome's great age. Similarly, in *Cynthias Reuells* (1600):

> the ladie ARETE, or *vertue*, a poore *Nymph* of CYNTHIAS traine, that's scarce able to buy her selfe a gowne

is, in fact, a powerful chastiser of prodigality and corruption. By a connotative slide, therefore, 'poor' is made synonymous with 'pure'.

Jonson is noted for the hard precision of his images; and justly. He builds magnificently so that he may destroy:

> we will eate our mullets,
> Sous'd in high-countrey wines, sup phesants egges,

> And haue our cockles, boild in siluer shells,
> Our shrimps to swim againe, as when they liu'd,
> In a rare butter, made of dolphins milke,
> Whose creame do's looke like opalls . . .

The moral pleasure here stems from Mammon's lavishing of such imaginative effort on shrimps; and 'rare butter' is a good, serious joke. Jonson's recreation of the bulk and weight of corruption, of criminal-farcical expenditure, is superb. On the other hand, the persuasive weight of such imagery is sufficient to commit the listener to the acceptance of a good deal of simple and evocative language, as part of Jonson's vision of the good life. In the First Chorus of *Catiline*, the pejorative vision of the effeminate men of new Rome (a terse set of epithets: 'kemb'd', 'bath'd', 'rub'd', 'trim'd', 'sleek'd') works into a passage of much more tenuous verbal gesturing when old Rome is evoked:

> Hence comes that wild, and vast expence,
> That hath enforc'd *Romes* vertue, thence,
> Which simple pouerty first made . . .

And Fulvia's reply to Sempronia is also a case in point. Her:

> 'Twas vertue onely, at first, made all men noble

is plainly on the right side of the moral fence. 'All men' sounds very liberal and fine; so does the easy juxtaposition of 'vertue' and 'noble'. In practice, though, Jonson's social vision is as far from a Leveller's dream as it is from any real faith in a natural aristocracy. Phrases like 'all men' merely make gestures towards a proper definition.

Such evasions are to be distinguished from the valid workings of dramatic and rhetorical persuasion. Take, for contrast, this sequence from *Catiline*:

> CHOR[US].   The voice of CATO is the voice of *Rome*.
> CATO.   The voice of *Rome* is the consent of heauen!
> And that hath plac'd thee, CICERO, at the helme . . .

Here, the anadiplosis, linking Cato's word to the authority of heaven, appears as a legitimate rhetorical parable of the great chain of Being. And, in this context, the commonplace of ship-of-state (as, elsewhere, the commonplace for body-politic) appears as an authentic, even though limited, statement of civic faith. Jonson *is* pedantic, but his pedantry has little or nothing in common with the supposedly neurotic crabbedness of the unworldly scholar. He is pedantic as Fulke Greville in his *Life of Sidney* is pedantic: each is prepared to risk appearing over-scrupulous in the attempt to define true goodness:

> he piercing into mens counsels, and ends, not by their words, oathes or complements, all barren in that age, but by fathoming their hearts, and powers, by their deeds, and found no wisdome where he found no courage, nor courage without wisdome, nor either without honesty and truth.

In Jonson, too, the insistent comparison, qualification and paradox – sometimes appearing in-grown – is made in the teeth of a barren age. The world of the Roman plays, like that of many of the comedies, is a world full of false witness:

> We haue wealth,
> Fortune and ease, and then their stock, to spend on,
> Of name, for vertue . . .

If 'name' is a mere commodity, an advertisement without relation to substance or property or ethic, how can 'name' alone be trusted? 'Name' must be demonstrated as belonging or not-belonging to 'thing'; its properties plainly discussed. Hence, in Jonson's dramatic rhetoric, antithetical pairings are frequent: 'inward' opposes 'outward'; 'justice', 'law'; 'private gain' is won by 'public spoil'. In the world of the Roman plays, this degree of moral scrutiny is painful, but necessary. This is the celebration of a good man, the dead Germanicus, in *Sejanus*:

> He was a man most like to vertue'; In all,
> And euery action, neerer to the gods,
> Then men, in nature; of a body' as faire
> As was his mind; and no lesse reuerend
> In face, then fame: He could so vse his state,
> Temp'ring his greatnesse, with his grauitie,
> As it auoyded all selfe-loue in him,
> And spight in others.

Here, the heavy antitheses are wrenched across the line, with considerable moral and muscular effort. They do not fall harmoniously within the line-period – and within the limits of their own predetermined world – as do Dryden's fluent couplets to the good man, Barzillai:

> In this short File *Barzillai* first appears;
> *Barzillai* crown'd with Honour and with Years . . .

52

> In Exile with his Godlike Prince he Mourn'd;
> For him he Suffer'd, and with him Return'd.
> The Court he practis'd, not the Courtier's art:
> Large was his Wealth, but larger was his Heart:
> Which, well the Noblest Objects knew to choose,
> The Fighting Warriour, and Recording Muse.

In terms of Dryden's bland wit, Charles is unquestionably 'godlike'. This is the very point that Jonson labours to verify in the portrait of Germanicus: 'In all,/And every action, neerer to the gods, /Then men ...' Jonson's qualifications worry the verse into dogs-teeth of virtuous self-mistrust. If this passage is painfully spiky, Dryden's approaches fatty degeneration. 'For him ... with him', 'Large was ... larger was' are syntactically flabby, embodying the elevated complacency of the thought. The repeated '*Barzillai*... *Barzillai*' gestures in the general direction of virtue and goodwill. The overall tone of the Barzillai panegyric, it can be argued, has much to do with Dryden's thoroughgoing acceptance of his lot as laureate of the Tory oligarchy. The famously smooth irony of the opening of *Absalom and Achitophel* is, ultimately, an indulgent cherishing irony, making such a smiling gentlemanly thing of equivocation that only a puritan cit could object.

Jonson's sense of the complexity of Order is very different, both poetically and politically, from Dryden's practical political attack and defence. Jonson's awareness can produce both the richly suggestive ambiguities of the Catiline-Aurelia courtship, or Latiaris' speech about Tiberius, and the fine assurance of the Pembroke Dedication. But there are occasions when, it seems, Jonson succumbs to the contradictions of the age; when he is unable, even reluctant, to tie down and tame the airy, floating connotations in words such as 'noble', or 'vertue', or 'poverty'. The *Catiline* Chorus, for example, speaks of 'simple poverty' as though this, and this alone, accounted for the 'vertue' of Old Rome. A moral slogan such as this is as blandly shallow as anything produced by Dryden's political partisanship. And in *The New Inne* (1631) there is this, equally emotive, exchange:

> NUR[SE].   Is pouerty a vice? BEA[UFORT]. Th'age counts it so.
> NUR.   God helpe your Lordship, and your peeres that think so,
> If any be: if not, God blesse them all,
> And helpe the number o'the vertuous,
> If pouerty be a crime ...

Yet, in *Catiline* it is made explicit that 'need' is the reason that so many flock to the conspirators' support. And Volturtius, a turncoat and minor villain, is contemptuously dismissed by Cato without punishment; is even promised:

>   money, thou need'st it.
> 'Twill keep thee honest: want made thee a knaue.

Poverty, then, is a virtue only when it is associated with a remote, lyrically evoked agrarian order, a world so distant that it appears static; or when it provides a temporary refuge for missing persons of good family. The poverty that really preoccupies Jonson is the reverse of simple and is far from static; is, in fact, a prime mover in that constant flux of demand and supply; the seventeenth-century urban predicament.

This world of savage indecorum, which, like Tiberius, 'acts its tragedies with a comic face', which to Donne appears 'bent awrie', is a world that Jonson struggles to subdue. It may be true that:

> In *Catiline* as in *Sejanus*, Jonson appears wholly inaccessible to the attraction of the profound 'humanity' and psychology of Shakespearian tragedy.

but it is equally true that he is inaccessible to that periodic debauching which the Shakespearian 'humanity' suffered at the hands of, say, John Webster. There is nothing in Jonson's Roman tragedies to equal the celebrated 'syrup' speech in the Duchess of Malfi's death scene; and we should at least be grateful for this. One ought to be sceptical of 'timeless moments' in any art. Jonson redeems what he can:

>   It is not manly to take ioy, or pride
>   In humane errours (wee doe all ill things,
>   They doe 'hem worst that loue 'hem, and dwell there,
>   Till the plague comes) The few that haue the seeds
>   Of goodnesse left, will sooner make their way
>   To a true life, by shame, then punishment.

FOUR

'The True Conduct of Human Judgment': Some Observations on *Cymbeline*

When Queen Elizabeth was dead they cut up her best clothes to make costumes for Anne of Denmark's masques. These performances,

> especially after the advent of Inigo Jones in 1607, became the rage of Jacobean London. The Vision in *Cymbeline* was clearly designed in response to this taste . . .

Shakespeare's actual working brief is unfortunately not so clear. One cannot conclude that *Cymbeline* was created for the Court or even specifically for the Blackfriars. Evidence, which is extremely tenuous, seems to indicate a performance at the Globe, *c.*1611. It is of course true that, even if the play was designed primarily for the public stage, its author was a servant of the royal household 'taking rank between the Gentlemen and Yeomen'. The king, though not a niggardly patron, was easily bored and preferred hunting. On the other hand, if the King's Men were employed as purveyors of *divertimenti*, they were no worse off than, say, Monteverdi at Mantua and probably more highly favoured. In terms of the dignity of art this is not saying much but 'since princes will have such things' it is possible for serious artists to elicit a private freedom from the fact of not being received as they deserve.

It has been suggested that *Cymbeline* was a 'dual purpose' play, adaptable to either public or private production. It may, in that case, have said different things to different audiences. A Court or Blackfriars audience could have seen what it expected, 'an unlikely story . . . an exhibition of tricks'. In daylight, at the Globe, the 'tricks' just might have appeared as critiques, oblique metaphors or moral emblems. The supposition that a private audience was necessarily more sophisticated than a public one is questionable. There is an edgy watchfulness to the play's virtuosity which might indicate Shakespeare's reluctance 'to commit himself wholly to the claims of his material'. No such reluctance need be deduced. Even so, an element of reserve about the claims of orthodox mystique, or about the eloquence of current political mythology, might be demonstrable. Imogen sees Britain's relation to the world's volume as being 'of it, but not in't' (III. iv. 141), an enigmatic phrase which might equally well describe *Cymbeline's* capacity for private nuance in its unfolding of the supreme theme of national regeneration and destiny.

Myths were things of utility to Tudor and Stuart politicians. They were also, though more sensitively, things of utility to the dramatist. The thought of Shakespeare hamstrung and humiliated by his own sycophancy is obtrusive but not inescapable. In the light of Emrys Jones's persuasive essay it would be unwise to ignore the relevance of Stuart myth to our understanding of this play. Henry VII was the saviour of his people; his great-great-grandson James VI and I, the unifier and pacifier, was 'the fulfilment of the oldest prophecies of the British people'. To the medieval chroniclers Kymbeline was a man of peace whose reign coincided with the birth of Christ. Shakespeare could have read this in Holinshed and sensed the connotations. Milford Haven, where the secular redeemer had landed in 1485, was a hallowed place celebrated in Tudor and Stuart patriotic verse. Such numinous power is not unknown even in the twentieth century. A modern author voices his emotions about Stalingrad:

> I felt that Stalingrad was in the middle of the world, a place where the final conflict between good and evil was fought out . . .

Holy and awesome places do exist, and there is propriety in Imogen's being 'magnetized to this, enchanted, spot', 'this same blessed Milford' (III. ii. 60), and in the contrivance of the play's denouement there. It is less easy to concede, however, that the political myth provides

> a kind of interpretative key to events on the stage which, without such a key, appear insufficiently motivated, almost incoherent.

To suggest this while concluding that the play is finally guilty of 'a central fumbling, a betrayal of logic' is to say in effect that one still finds it incoherent. The symptoms have merely been associated with a different cause, with over-caution induced by its 'being too close to its royal audience'.

One's own case would be put rather differently. It would be that Shakespeare's involvement with the claims of his material is full and unqualified while, in his view of extraneous and imposed value-judgments, he remains singularly open-minded. The play's virtuosity is manifested in the association of committed technique with uncommitted observation. It is aware of the fact of compromise but it is in no sense compromised. Recognizing the aims and expectations of such a company as the King's

Men, who were undoubtedly popular at the Jacobean Court, one also admits the reasonableness of the suggestion that Shakespeare may have had James in mind when writing *Cymbeline*. Scepticism should be directed not so much against an extremely plausible hypothesis as against a number of moral and aesthetic conclusions which have been drawn from it. Even if it were an established fact that a royal audience actually attended a performance of this play, the proposition that Shakespeare therefore 'officiously' protected his royal protagonist from 'the consequences of his weak nature and ill-judged actions' would still be open to debate. It could be argued that *Cymbeline* is more subtly and sensitively handled than the objection allows and that the play shows considerable openness in places where its metaphysics have been assumed to be nothing if not transcendental. One of these is the love of those 'two supremely excellent human beings' Imogen and Posthumus; the other is the theme of the king's peace. Since one accepts that the strands of personal and national destiny are interwoven, to debate these elements is to study the crucial implications of the play. In each case, one would suggest, Shakespeare stands back not through timidity or unconcern but in order to obtain focus. His concern is with nearness and distance, with adjustment of perspective; in his handling of character and situation he makes a proper demonstration of that concern.

In recent years it has begun to be possible to understand Imogen in terms proper to the dramatic context. For so long, and for so many admirers, she had been simply 'one of the great women of Shakespeare or the world'. There were, of course, dissenting voices. Shaw's reaction to the adulatory tradition was rough but not impertinent:

> All I can extract from the artificialities of the play is a double image – a real woman *divined* by Shakespear without his knowing it clearly . . . and an idiotic paragon of virtue produced by Shakespear's *views* of what a woman ought to be . . .

The antithesis is arbitrary, the premiss invalid, the naturalistic criterion misleading. Shaw supposes the dramatist to be merely intuitive when he is, in fact, quite deliberate. Shakespeare has shaped his play to procure the reality of the woman from the romance of her setting. Shaw's comments have considerable value however. He perceives a dichotomy and his suggestion of a double image is very relevant. It could be said that *Cymbeline* involves the realization of double images, not as Donne's visualization of the spirit of mutual regard ('My face in thine eye, thine in mine appeares') but as we might refer to double exposures in photography: accidental or contrived palimpsests that come from one view having been superimposed on another. Does the supremely excellent Imogen commit a breach of natural propriety

57

when she speaks of Posthumus, after his departure, to his servant Pisanio?
She wishes that she had made him

> swear
> The shes of Italy should not betray
> Mine interest, and his honour. (I. iv. 28–30)

These words are said to be 'quite out of character' but the objection rests on
a preconceived notion of what this character comprises. Imogen's remark is
not a happy one and it is difficult to see how, in the circumstances, it could
be. Shakespeare's interest is in certain kinds of immaturity and in the
inaccuracy and imbalance affecting relationships. Imogen wishes, when she
meets 'Polydore' and 'Cadwal', that

> they
> Had been my father's sons, then had my prize
> Been less, and so more equal ballasting
> To thee, Posthumus. (III. vi. 76–8)

'Ballasting' could mean 'freight' or 'stabilizing weight'. Imogen seems to
mean that a male heir to the throne would have reduced her own status to a
level nearer that of Posthumus. Her words acknowledge how Posthumus'
conquest of her must have appeared to some observers – as a piratical act –
and imply a fear that those who foresaw disaster in such inequality may have
been right. In the play's opening scenes the breach of convention and the
strained court relations create the original milieu for inflationary panegyric
and protestation. The love-cries of Posthumus and Imogen are a lyrical
defiance of circumstance and a breeding-ground of error. The courtiers
involve themselves in partisan polemic. A 'perfect' Posthumus is necessary
to their own self-respect. What they say of him is bound to seem 'too good to
be true' but to suppose that Shakespeare is himself off-balance is to miss the
point of the situation. It has been suggested that 'Shakespeare intends us to
accept the First Gentleman's estimate of [Posthumus'] virtue'. That
Shakespeare intends nothing of the kind has been argued as strongly. The
wager-plot reveals

> beneath Posthumus' apparently perfect gestures an essential meanness in
> the man himself and in the conventional virtue that he embodies.

This is perhaps overstressed. Meanness is not intrinsic to Posthumus'
nature and can be eradicated. Nevertheless, the point is taken. He survives
and matures in grace, but from a condition in which he was possibly more

like Iachimo or the despised Cloten than his admirers could have credited. Iachimo, at least in the wager-scene, is not so much a villain as a catalyst for Posthumus' own arrogance and folly. At the end of the play both men express personal remorse in terms that bear comparison. Iachimo feels that guilt 'takes off [his] manhood', the air of Imogen's homeland 'enfeebles' him (v. ii. 2–4). Posthumus' conscience is 'fetter'd' (v. iv. 8). One man regards his honours as but 'scorn' (v. ii. 7), the other thinks of himself as debased coin (v. iv. 25). Posthumus, in forgiving Iachimo, is in a sense forgiving himself. Implied comparisons between Posthumus and Cloten are also to be discerned. The suggestion has indeed been made that 'the two characters, never on stage together, should be played by a single actor' and the significant visual pun, whereby Imogen is convinced that Cloten's body is her husband's, has frequently been noted. It is more significant that Posthumus is not Cloten and one sees that the irony reflects as strongly upon Imogen as upon her husband. What kind of man did she imagine him to be? Did she truly know him at all? The double irony is beautifully released in her mistaken tears.

At such points Shakespeare is treating ideas of false assumption, false connection. Dr Leavis's objection that Posthumus' jealousy 'has no significance in relation to any radical theme, or total effect, of the play' seems without real foundation. When the husband recoils from his wife's supposed infidelity the play focuses, not in any simple way upon character, but upon the experience of relationship:

> Me of my lawful pleasure she restrain'd,
> And pray'd me oft forbearance: did it with
> A pudency so rosy, the sweet view on't
> Might well have warm'd old Saturn; that I thought her
> As chaste as unsunn'd snow. (II. v. 9–13)

Mr Traversi rightly points to the 'sensual poison' at work here. Are we, though, to regard this merely as the extravagance of Posthumus' contaminated imagination? The paragon of romantic love now sees his marital function, somewhat obtusely, as the taking of 'lawful pleasure', yet his words, unlike those of Claudio or Leontes, do seem to imply a degree of genuine recollection rather than of purely fanciful diatribe. The recollection is of a somewhat vulnerable awkwardness on both sides. These are, after all, wedded lovers and not two virginal puppets from *The Faithful Shepherdess*, to whose theoretical exacerbations phrases like 'A pudency so rosy' and 'As chaste as unsunn'd snow' seem more properly to belong. In spite of what has been called 'the puritanical emphasis on pre-nuptial purity in *The Winter's Tale* and *The Tempest*' the tone of Perdita's 'But quick, and in mine arms' seems the real heart of innocence whereas this admittedly second-hand description of Imogen's nuptial modesty does not. Conversely, Imogen and Posthumus have, from the opening of the plot, a grim actuality of separation to contend with; a matter that other young lovers in the Romances do not have to face so blankly. The very fact that they are already married when the play begins is what makes them exceptional. They are even then beyond the conventional romantic consummation. As Imogen imagines her husband's ship sailing further and further away and the figure of Posthumus becoming smaller and smaller, the language with which she tries to grasp her loss has a touch of that extravagance displayed by the courtiers who try to force Posthumus into attitudes of perfection. We are not unmoved by the human credibility of this but we should recognize that, between such over-refinement of invention and the bitter grossness later displayed by Posthumus, there is a moral and emotional hiatus. It is true that there is no 'deep centre' here, because there is loss or failure of contact. This would seem to be, more than anything else, evidence of Shakespeare's psychological accuracy.

The state of the relationship here is a state of tactlessness, meaning 'the absence of the keen faculty of perception or discrimination' or, more simply, 'being out of touch'. There is a vice which, according to Bacon,

> brancheth itself into two sorts; delight in deceiving, and aptness to be deceived; imposture and credulity; which, although they appear to be of a diverse nature, the one seeming to proceed of cunning, and the other of simplicity, yet certainly they do for the most part concur . . . as we see it in fame, that he that will easily believe rumours will as easily augment rumours . . .

A feasible corollary might be that there is a kind of naïvety which asks to be devoured and a natural and partly unconscious collusion between the deceived and the deceiver: between, for instance, Posthumus and Iachimo, Imogen and Iachimo, Cymbeline and the Queen. There is a language surrounding this collusion, a twofold idiom of inflation, on the one hand fairly innocent and spontaneous, on the other a matter of policy. Of the First Gentleman's gush of praise for Posthumus' qualities it could be said, in a phrase used of Shakespeare's royal master, that 'the sums he gave away so easily were not his to give'. This adds some point to Posthumus' later reference to himself as 'light' coin. As though sensing the folly of former inflation, he recoils to the other extreme of debasement.

The question of the play's rhetorical patriotism is more problematical. Cloten's 'tactlessness', the juxtaposition in him of oafish petulant inadequacy and outspoken nationalistic fervour, has been discussed by Wilson

Knight. There is possibly a slight oversimplification in the description of Cloten's rudeness to the Roman envoy as

> British toughness and . . . integrity . . . Cloten . . . is for once in his element without being obnoxious.

One would suggest that, on the contrary, Cloten is obnoxious, and so is the Queen; their patriotism is an exacerbation rather than an amelioration of this. It is somewhat reminiscent of the old ranting of *Locrine*. The fact that a proportion of Shakespeare's audience would undoubtedly identify itself with Cloten's xenophobia and thus be trapped in an emotional cul-de-sac makes for an interesting piece of dramatic ambivalence. Television audiences suffer the same kind of dilemma when torn apart by the antics of Mr Alf Garnett.

> The Britons defy Troglodytes, Aethiopians, Amazons and 'all the hosts of Barbarian lands', if these 'should dare to enter in our little world'.

Finally, Cymbeline admits that his resistance to Rome had been instigated by his 'wicked queen' (v. v. 463). In view of James's self-appointed task as European peacemaker, Shakespeare's portrayal of such erroneous and ferocious insularity might be regarded as a tenable risk; but the question is boldly treated and taken further than mere diplomacy would have required.

There is, in fact, little indication of any desire on the part of the dramatist to make discreet alignments with official Jacobean policy, and the suggestion that 'the character of Cymbeline . . . has a direct reference to James I', in so far as this implies a commitment to eulogy, is not convincing. One would accept, however, that Shakespeare caught, accurately and retentively, a certain tone of Jacobean mystique and that an oblique awareness of royalist views and demands can be detected. It is as though, in the play's finale, we were presented with a completely open situation. A British king is seen for what he is: uxorious, irrational, violent when prodded, indulgent, of absolute status and ultimately invulnerable. This might be taken as a disinterested view of the 'law' of Prerogative. The concluding atmosphere of transcendental peace has been found appealing. The king reiterates the word 'peace' three times in eight lines (v. v. 478–85) and also speaks of 'my peace' (v. v. 459). One is reminded that

> some earlier editors were so offended by the apparent megalomania of '*my* peace' that they emended to 'by'.

The editorial procedure may have been questionable but the original suspicion may not have been entirely groundless. Cymbeline's evocation of peace is a happy combination of the cursory and the opulent. To his daughter he admits his 'folly' (v. v. 67); politically, the guilt is laid upon a defunct secondary cause. There is a passage in *The Trew Law of Free Monarchies* (1603) which simultaneously concedes and vetoes possible monarchical deficiencies. The people, writes James, should look to the king:

> fearing him as their Iudge, louing him as their father; praying for him as their protectour; for his continuance, if he be good; for his amendement, if he be wicked; following and obeying his lawfull commands, eschewing and flying his fury in his vnlawfull, without resistance, but by sobbes and teares to God . . .

Such a thesis, many-faceted and unanswerable, destroys objections by assimilation. Critics who object to the plodding inadequacy of the verse spoken by the 'poor ghosts' of the Leonati (v. iv) may not have considered that the play needs an element of formal pleading which is quite distinct from the eloquent magnanimity of confirmed majesty. Their wooden, archaic clichés are like an emblem of old and rather weary sincerity, whereas Cymbeline's concluding oratory is the reaffirmation of the mystique of status.

Shakespeare is in accord with the *Trew Law* to the extent that he brings the play to an end 'without resistance'. It is neither adulatory nor satiric, but observes what is there. Altogether, one feels that *Cymbeline* betrays nothing, is eminently logical and does not fumble. To be left, at the end, with things inexplicable and intractable is a perennial hazard for all artists; but in Shakespeare's last plays an acceptance of this seems to be at the heart of his dramatic vision. It is hardly scepticism, rather a kind of pragmatism, a necessary counterpoise to the thoroughly pragmatic 'myths' of the Stuarts. It has been said that 'the ending of *The Tempest* is very moving, not least because it is so reticent'. Reticence about things that cannot be reconciled is a characteristic of the last plays. In *Henry VIII* this tacit recognition is itself reduced to a formula as the two conflicting obituaries of Wolsey are presented in sequence by Queen Katherine and Griffith. The play sounds a note of political quietism while in *Cymbeline* the tension of paradox is persistently felt. In each case, however, we are involved in a dualistic acceptance of things as they are. Such dualisms seem to avoid the chain of cause and effect which drives the tragedies. They avoid, also, the formal concept of Tragedy in which 'hamartia' indicates the irreparable severing of tragic experience from the normal conduct of life; the start of the irreversible slide down the scale of act and consequence. In the late plays hamartia appears as something integral to the human condition, innate, to be lived with.

Admittedly, this observation might be little more than a truism stretched awkwardly between time-serving cynicism, lazy nostalgia and altruistic resilience. There are comments of a relatively simple kind, spanning the range of Elizabethan-Jacobean chronology, which would qualify as any one of these. John Lyly argued, sometime in the 1580s, that

> If wee present a mingle-mangle, our fault is to be excused, because the whole worlde is become an Hodge-podge.

One finds a reminiscent explanation prefaced by the composer Thomas Tomkins to his *Songs* of 1622. These, he hopes, are

> suitable to the people of the world, wherein the rich and the poor, sound and lame, sad and fantastical, dwell together.

Uncomplicated observations of this nature are possibly imitated in Hamlet's

> the age is grown so picked that the toe of the peasant comes so near the heel of the courtier, he galls his kibe. (v. i. 151–3)

Such an aphorism is far from representing the mood caught in Shakespeare's so-called Romances. Their spirit is altogether more complex and has been excellently described. Even so, Lyly and Tomkins are not alone in their persuasion that the world of common observation is one which disobeys at every turn the world of overriding mythology and that such contradictions affect the nature of art and the ways in which people respond to it. A sense of mundane fallibility is stated with characteristic authority by Bacon. Of the false appearance imposed by words he writes:

> it must be confessed that it is not possible to divorce ourselves from these fallacies and false appearances, because they are inseparable from our nature and condition of life.

He adds, however, that 'the caution of them . . . doth extremely import the true conduct of human judgment', thereby converting what might have been a merely passive acceptance into something active and therapeutic. The last act of *Cymbeline* reveals a possibly analogous attitude, particularly as the play's major dilemma has been brought about largely by false appearances imposed by words. Just how far Shakespeare is from an indulgence in flaccid geniality is shown by the summary disposal of Cloten and the crazed death of the Queen. One might say of the final gallimaufry, mirroring in so many ways the sad and fantastical, that it avoids instruction, whether comic or tragic, but is, in Bacon's sense of the term, 'cautionary'.

It has been suggested that 'there is more Baconism in late Shakespeare than is normally recognized'. This is clearly a challenging suggestion; it would perhaps not be possible to establish 'Shakespeare's Bacon' on the same basis of documentary evidence as is available for Shakespeare's Plutarch or Shakespeare's Montaigne. It is arguable, however, that *Cymbeline* turns on an awareness of experiment or experiential knowledge. Imogen's cry:

> Experience, O, thou disprov'st report! (IV. ii. 34)

which is, in some senses the leitmotiv of the play, is uttered at a point where the threads of the action are interwoven: the marriage story with that of the lost children and the theme of Britain. Its facets illuminate all areas. 'Experimentation had long had its connections with magic', and the Queen who possesses, or imagines she possesses, 'strange ling'ring poisons' (I. v. 34), 'mortal mineral' (V. v. 50), is a practitioner of Bacon's 'degenerate Natural Magic'. Shakespeare twists and exaggerates one side of Bacon's casuistical dichotomy between lawful and unlawful investigation. Wilson Knight speaks of the Queen's 'instinctive' support of her son, a perceptive epithet suggesting that element of unregenerate energy which also inspires her murderous cunning and her evil exploitation of natural resources. The two lost princes are also seen, in some of their acts, as housing 'primal nature in its ferocious . . . aspect'. There is, however, a major difference between the Queen's nature and theirs. Guiderius kills Cloten with what seems an untroubled animal reflex but the princes' moral intelligence partakes of 'rehabilitated nature'. They are moved by that sense of proportion which is a requirement of true relationship. When Arviragus says:

> We have seen nothing:
> We are beastly: subtle as the fox for prey,
> Like warlike as the wolf for what we eat:
> Our valour is to chase what flies . . . (III. iii. 39–42)

this is not, as it might seem, emblematic fatalism. It was a Renaissance understanding that 'Man had learned about his own nature from his observance of animals' and Arviragus' self-reproach has an experiential emphasis. The scene provides a significant departure from orthodox commonplace. Belarius' truisms about Court *v.* Country are a point of departure rather than a place of rest and have dramatic relevance as revealing the force of an obsession rather than the truth of an ideal. His values are questioned by the two princes who argue, in effect, that *a priori* value must be submitted to, and substantiated by, experience.

Cymbeline has been called an experimental drama mainly because of its competitive awareness of the new Jacobean stage-fashion for Romance. The term could be applied in other ways, since the play is significantly experimental with its characters, values and situations. One simple essential in an experiment is time. Philario says of Posthumus, with unintentional irony:

> How worthy he is I will leave to appear hereafter, rather
> than story him in his own hearing.   (I. iv. 33–5)

The events of the play perform what the First Gentleman admitted he could not: they 'delve' Posthumus 'to the root' (I. i. 28). To say that the realization of character involves the realization of situation is to proclaim the obvious. Such an emphasis is necessary, however, for *Cymbeline* is above all a study of situation, relationship, environment and climate of opinion. If we are 'on occasion . . . content to forget the play and concentrate on its heroine' very great harm is in fact done. Again, the question of possible dramatic bias may be raised. If Cymbeline himself, who is both character and 'climate', seems to elude the extremes of scrutiny one need not infer evasiveness on the dramatist's part. The king's decision, though victorious, to pay his 'wonted tribute' after all (v. v. 462) invites us to take it seriously, as a gesture of deep humility and the play's moment of truth. It might be taken more appropriately as a token of considerable ambivalence. Its solemnity exists for the participants in the immediate situation while the play itself contains a reserve of irony. Cymbeline's magnanimous gesture is also one of Shakespeare's supremely comic moments. If it is a moment of truth this is because it lays bare the absurdity of the original mouthing and posturing and pointless antagonism.

> Laud we the gods,
> And let our crooked smokes climb to their nostrils
> From our blest altars.   (v. v. 476–8)

The poetry is both plangent and jagged. 'Should not the smoke of an acceptable sacrifice rise undeviously to the heavens?' It is a proper question. Cymbeline's command makes an uncontrollable element appear deliberate, converting the accidental and the thwart into a myth of order and direction. Such a myth reflects a royal sense of occasion and the mystique of status but one recognizes that the phrase has the kind of resonance which could bring more distant connotations into sympathetic vibration. This evocation of majestic aplomb belongs, as it were, partly to Cymbeline the 'character', partly to Shakespeare the dramatist: on the one hand 'business as usual', on the other 'this is what the reality of Prerogative is like'. Such a double-take would be quite in accord with the play's feeling for compromise. *Cymbeline's* sense of finality, seen in this light, resides in its capacity to annul, through time, exhaustion and sleep, the business of the wicked and arrogant and the impetuosity of the immature. 'Crooked smokes', considered as the dramatist's 'own' metaphor, is like silent music, a visual rendering of the favourite Jacobean musical device known as 'chromatic tunes'. What this means is embodied in a line of Pisanio's:

> Wherein I am false, I am honest; not true, to be true.   (IV. iii. 42)

This is the taming of 'false relation' to a new constructive purpose. Dissonance is the servant preparing the return of harmony. In the words of John Danyel's lute-song of 1606:

> Uncertain certain turns, of thoughts forecast,
> Bring back the same, then die and dying last.

Cymbeline, which 'aims at effecting the gratification of expectancy rather than the shock of surprise', brings back the same in both grace and mediocrity. It brings back husband to wife and tribute to Rome. It also reaffirms the king in the dualism of his selfhood and his prerogative. Shakespeare is perhaps ready to accept a vision of actual power at cross-purposes with the vision of power-in-grace; 'the real world', in fact, 'in which the life of the spirit is at all points compromised'. To make such a suggestion is not to push Shakespeare into placid cynicism or angry satire but it is to feel that he at least knew the difference between acceptance and indulgence. *Cymbeline* has been called 'enchanted ground'. If it is, then that ground is

> drenched in flesh and blood, Civil History, Morality, Policy, about the which men's affections, praises, fortunes, do turn and are conversant . . .

steeped, that is, in those knowledges which men generally 'taste well'.

FIVE

Jonathan Swift:
the Poetry of 'Reaction'

In general, when we use the word 'reaction', we may be referring to a supposedly 'retrograde tendency' in politics or to a 'revulsion' of feeling. Since the political and pathological aspects of Swift's work have been amply debated, the purpose of the present exercise may appear open to question. It should be emphasized that, in proposing the term 'reaction' for this discussion, my chief endeavour is to define and describe an essential quality of Swift's creative intelligence: the capacity to be at once resistant and reciprocal.

I am aware that, in the consideration of Augustan poetry in general and Swift's poetry in particular, arbitrary assumptions as to the irreconcilability of wit and feeling still carry weight; and Swift is especially vulnerable to those distortions of interpretation which occur when ideas are extracted from the texture of language. To state my own case in its most basic terms is to say that if, as a moralist, Swift's concern is with the ordering of acceptance and rejection, as a moral artist he can transfigure his patterns. We may readily perceive that this demands from the poet a most sensitive awareness of the things which are being reacted against and that in his finest work this sensitivity works as a catalyst to transform, say, autocratic disdain into a cherishing particularity. We recognize just as readily that this awareness can be fully realized only in the medium of language itself, the true marriage, in words, of wit and feeling.

*　　*　　*

It is no longer possible to explain Swift's genius in terms of convenient myth. That he was not 'mad' is attested by eminent medical and scholarly authority; that he suffered, in 1742, a brain-lesion followed by senile decay seems probable and, in view of his advanced age, unremarkable. Nevertheless, for a considerable part of his life he was certainly ill and demonstrably eccentric, concerned to the point of obsession with manifestations of the absurd and irrational in himself, in particular enemies and friends, and in society at large. Certain episodes in his career would have justified such concern, among them his fruitless efforts to reconcile the quarrel between

Bolingbroke and Oxford, the oppressive infatuation of Esther Vanhomrigh, the threat of arrest by the Walpole administration and of physical assault by Richard Bettesworth. These were matters of more than clinical observation, but such observations, in the form of philosophical and political theory, were not overlooked. Swift read and annotated his volumes of Machiavelli, Bacon, Harrington and Hobbes. Sir William Temple had been attracted by the truism that events in history could be largely explained by noting the 'passions and personal dispositions' which rule the private lives of public men; and the master's fondness for the commonplace seems to have been shared by his pupil. One may feel, therefore, that to view the intimate eccentricities of Swift in their larger context is in accord with his own creative disposition. Such poems as 'An Apology to the Lady Carteret', 'Lady Acheson Weary of the Dean', and 'Verses on the Death of Dr Swift' suggest that, as a poet, he could represent personal predicaments emblematically and turn private crisis into public example. W. B. Yeats, whose view was understandably influenced by the prevailing myth of Swift as *poète maudit*, to some extent justified his acceptance of the legendary figure by making it representative of the destruction of an epoch. The final silencing of Swift was the silencing of the voice of moral authority in his place and time: 'more than the "great Ministers" had gone'.

In Swift a sense of tradition and community is challenged by a strong feeling for the anarchic and the predatory. A necessary qualification is that the appeal of Community exists not as a fine Platonic idea but as something soberly lived, taken into the daily pattern of conduct and work. A reader of his correspondence, as of the *Journal* and birthday poems to Stella, comes to accept the real presence, as well as the ritual, of his friendships. This affects the poetry, in particular its power to move with fluent rapidity from private to public utterance and from the formal to the intimate in the space of a few lines. At times, in the letters to Bolingbroke and Oxford, what is private is simultaneously public in its implications. It is of course true that when one has a few good friends and those friends happen to be the most important men in England, E. M. Forster's injunction 'only connect' has a particularly happy significance. It might be objected that such ideal conditions hardly existed for Swift after the debacle of 1714; it might also be remarked that the familiar verse which, we are told, 'occupied so much of Dean Swift's leisure' in Ireland became a half-defensive exaggeration of the sudden failure to connect, a play on the vicissitudes of being 'in' or 'out':

> When I saw you to-day, as I went with Lord Anglesey,
> Lord! said I, who's that parson? how awkwardly dangles he!
> When whip you trot up, without minding your betters,
> To the very coach side, and threaten your letters.

This is the conversion to familiar nuance of the public admonitory tone of seventeenth-century satire, but for the intimate joke to have bite it still needs the public contact, an edge of hard fact. In a letter to Pope, years afterwards, Swift wrote that:

> I was t'other day reckoning twenty Seven great Ministers or men of Wit and Learning, who are all dead, and all of my Acquaintance within twenty years past . . .

and continued, as though trying to balance his account, with the names of nineteen 'men of distinction and my friends who are yet alive'. Primarily an ageing man's game of patience, it is still a suggestive catalogue both for the timbre of the voice and for the data it provides. Several of those named, including Lord Berkeley, Lord Carteret, the Duke of Ormonde and the Earl of Pembroke had held high office as Lord Justice or Lord Lieutenant in Ireland. The younger Swift, 'taking his ease at the Castle', had enjoyed friendly acquaintance with Berkeley and Pembroke. Carteret's term of office (1724–30) enabled the old and celebrated writer to exercise a nice intricacy of formal yet familiar association with a man whose intellect was at once engaged and frustrated in the practicalities of government. However daunting the periods of Irish reclusion may have seemed, they were far from bringing total eclipse.

Swift has been called 'a practical politician in everything he wrote'; he was, in a variety of senses, a born administrator. Dr Johnson, who was not over-indulgent towards Swift's memory, admitted his capability as Dean of St Patrick's. Professor Ehrenpreis has remarked that, in Ireland, the administrative class to which Swift's family belonged could be 'numbered in the hundreds' yet formed 'a kind of social capital on which he drew interest for the rest of his life'. It was necessarily a self-contained society. Ireland's ruling caste, as the historian William Lecky pointed out, was 'planted in the midst of a hostile and subjugated population'.

Among the coterie of Swift's Dublin years were a number of people, active in their public lives, who wrote occasional verses and sometimes took part in humorous rhyming contests with the Dean and other friends. His literary acquaintance included Dr Patrick Delany, Dr Richard Helsham, George Rochfort, William Dunkin and the Earl of Orrery. Stella also wrote, and a character in Yeats's *The Words Upon the Window Pane* claims that she was a better poet than Swift; though on the evidence of the three pieces which are attributed to her the argument is untenable. These verses are less impressive than some by Swift's other close friend, Dr Thomas Sheridan. 'Tom Punsibi's Letter to Dean Swift' (*c.*1727) is an immaculate example of that style of familiar naturalism in which Swift himself excelled:

When to my House you come dear Dean,
Your humble Friend to entertain,
Thro' Dirt and Mire, along the Street,
You find no Scraper for your Feet:
At this, you storm, and stamp, and swell,
Which serves to clean your Feet as well:
By steps ascending to the Hall,
All torn to rags with Boys and Ball.
Fragments of Lime about the Floor,
A sad uneasy Parlor Door . . .

Sheridan's description of the miry street may be an admiring echo of Swift's earlier 'A Description of a City Shower' (1710) but it is clear from the accuracy of his writing that Swift had found a reciprocating talent. Apparently:

> so successful was Sheridan in imitating Swift's style that even Delany found it difficult to distinguish the work of one from that of the other.

But if 'Tom Punsibi's Letter' is imitation, it is so in the good neo-classical sense and is distinct from simple mimicry or mere plagiarism. Familiarity with Sheridan's work may have given Swift a much-needed incentive for his own familiar verse, a more racy and immediate contact than the admired Pope and a disciple and confederate in the 'Hudibrastic' line.

Swift, in his old age, is reputed to have had the whole of Butler's *Hudibras* by heart and it is somewhat surprising to find no mention of this work in the catalogue of his library. Swift's library was admittedly not widely representative of the achievements of modern English poetry. He had Spenser, Milton, Pope, Prior, Gay and Parnell but not, apparently, those two predecessors with whom, as a satirist, he is now most often associated. Of these, Butler is one, and for the absence of Rochester there may have been sufficient reason since even the libertine Earl's return to Christianity had been made under the aegis of the detested latitudinarian Bishop Burnet. Although Swift 'never was one of his Admirers' it is difficult not to be in sympathy with those critics who cite the 'agony and indignation' of Rochester's major satires as a precedent for Swift's own work in the genre and hard not to suppose that the Dean had read them, perhaps with mixed feelings about their 'Hobbesian' views but with a professional attention to detail. Compare Rochester's

> With mouth screwed up, conceited winking eyes,
> And breasts thrust forward, 'Lord, sir!' she replies.

70

'It is your goodness, and not my deserts,
Which makes you show this learning, wit, and parts.'
He, puzzled, bites his nail, both to display
The sparkling ring, and think what next to say,
And thus breaks forth afresh: 'Madam, egad!
Your luck at cards last night was very bad:
At cribbage fifty-nine, and the next show
To make the game, and yet to want those two.

with Swift's

'Stand further Girl, or get you gone,
'I always lose when you look on.
Lord, Madam, you have lost Codill;
I never saw you play so ill.
'Nay, Madam, give me leave to say
'Twas you that threw the Game away;
'When Lady *Tricksy* play'd a Four,
'You took it with a Matadore.

That the advantage of this comparison appears to be with Rochester is not wholly due to the exigencies of arbitrary selection. He is a lyric dramatist: the 'mouth screwed up', the gesture with the ring, seem to occur in a sequence of human cause and effect. Rochester peoples a situation with actors who themselves grasp that they are 'situated' and must therefore act to live. Swift, it might seem, though often praised for his sense of situation, shows less concern for human predicament than for the more obvious tones of social behaviour. He converts manners into mannerism; or he presents a formal setting which serves as a *pied-à-terre* for punitive excursions:

My female Friends, whose tender Hearts
Have better learn'd to act their Parts
Receive the News in *doleful Dumps*,
'The Dean is dead, (*and what is Trumps?*)'

Here, though, the rhyming wit itself works like a 'trump' or triumph to snatch brilliant personal success from a position of elected disadvantage. A constant preoccupation with verbal routines is a characteristic of Swift, as may be seen in *A Proposal for Correcting, Improving and Ascertaining the English Tongue* (1712) or *A Compleat Collection of Genteel and Ingenious Conversation* (1738), and his satiric art in 'Verses on the Death of Dr Swift' depends on a seemingly perfect rapport between clichés. The effect does not merely require the bathos of the move from 'dead' to 'Trumps'; it also needs a

71

constant frisson or sympathetic vibration between truisms, 'Friends', 'tender', 'doleful'. There is even a reverse thread of ironic suggestion linking 'News' and 'Friends'. The very solidity of Rochester's characterization is itself releasing. Swift's nuance-haunted repartee puts the burden of correct or incorrect response on the reader and is capable of employing pathos as a trap for obtuse decency:

> 'The Dean, if we believe Report,
> 'Was never ill receiv'd at Court:
> 'As for his Works in Verse and Prose,
> 'I own my self no Judge of those:
> 'Nor, can I tell what Criticks thought 'em;
> 'But, this I know, all People bought 'em;
> 'As with a moral View design'd
> 'To cure the Vices of Mankind:
> 'His Vein, ironically grave,
> 'Expos'd the Fool, and lash'd the Knave:

*   *   *

In order to isolate and describe Swift's own idiom it is necessary to recognize contacts, both those which excite and those which inhibit the growth of idiosyncratic art. If Swift's friendship with Thomas Sheridan stands at one pole of the creative field, Addison's influence possibly represents the opposite extreme. It is known that Addison drastically edited two of Swift's poems, 'Vanbrug's House' and 'Baucis and Philemon', by 'cutting out many strokes that gave vigour and force to the description'. Although Swift always referred to Addison in terms of personal respect it is clear that Swift's creative tact was a very different thing from Addison's literary taste. Nor was there always a perfect understanding with such close acquaintances as Delany and Orrery, who were sometimes disturbed by Swift's breaches of etiquette. It is not altogether astonishing to find in Swift's poetic satire a certain amount of irritation at the spurious proscriptions of false delicacy and a clear distinction between squeamishness and decorum. The question of 'reaction' implies more than a manipulation of antitheses. With many aspects of the consensus of taste Swift was undoubtedly able to agree, and it would be patronizing to suppose that he necessarily regarded himself as sacrificing original liberty on the altar of caste. Despite the possibly inhibiting surveillance of Addison or Temple and the sympathetic reservations of Delany and Orrery, Swift's poetry gained more than it lost by his overall adherence to the major canons of his class. 'The Journal of a Modern Lady' (1729) contains a scathing account of the trivial malice of social gossip:

> Nor do they trust their Tongue alone,
> To speak a Language of their own;
> But read a Nod, a Shrug, a Look,
> Far better than a printed Book;
> Convey a Libel in a Frown,
> And wink a Reputation down;
> Or by the tossing of the Fan,
> Describe the Lady and the Man.

But, setting ethics aside, this 'language' is no mean achievement in quick, economical mimicry. Swift never did leave ethics aside but it is still true that his poems have a good deal of the gadding energy and, at times, something of the spleen and complacency of the society which they inhabit.

It may be that close-knit communities need to evolve sophisticated weapons of control. Eighteenth-century verse sometimes aims to 'disarm'. *The Rape of the Lock* is such a poem; Swift's 'Cadenus and Vanessa' (1713) is another. This work originated in the Esther Vanhomrigh affair; Swift was 'embarrassed but unwilling to end the situation'. Ostensibly a tale in verse, 'Cadenus and Vanessa' is a working blueprint for the kind of poem which it then discovers itself to be; its task is to test various techniques against varying situations and successfully reduce a dangerous immediacy to a more remote hypothesis:

> She railly'd well, he always knew,
> Her Manner now was something new;
> And what she spoke was in an Air,
> As serious as a Tragick Play'r.
> But those who aim at Ridicule
> Shou'd fix upon some certain Rule,
> Which fairly hints they are in jest,
> Else he must enter his Protest:
> For, let a Man be ne'er so wise,
> He may be caught with sober Lies;
> A Science which he never taught,
> And, to be free, was dearly bought:
> For, take it in its proper Light,
> 'Tis just what Coxcombs call, *a Bite*.

Given the situation between Swift and Vanessa, such terms as 'in an Air', 'As serious as', 'may be caught', 'a Tragick Play'r' are a breath-taking defiance of gravity. The passage turns on the words 'railly'd' and '*Bite*', two technical terms out of the social rhetoric of the time. The requirements for a successful 'bite' were specified by Swift himself:

73

You must ask a bantering question, or tell some damned lie in a serious manner, and then she will answer or speak as if you were in earnest: and then cry you, 'Madam, there's a *bite*'.

Receiving an answer 'as if you were in earnest' suggests, in the Swift–Vanessa relationship, a denouement more far-reaching than the crisis of a rhetorical game; yet the whole point of the rhetoric is to defuse the emotional charge.

The somewhat complex question of raillery as an eighteenth-century social and literary phenomenon has been the subject of much able and detailed discussion. In terms of pure theory it is possible to distinguish two major types: 'ironical praise that is really satire or reproof' and its opposite, 'something that at first appeared a Reproach, or Reflection; but, by some Turn of Wit unexpected and surprising, ended always in a Compliment, and to the Advantage of the Person it was addressed to'. Swift favoured, in principle, the second type and was particularly careful to make the distinction between constructive raillery and mere abuse:

> So, the pert Dunces of Mankind
> Whene're they would be thought refin'd,
> Because the Diff'rence lyes abstruse
> 'Twixt Raillery and gross Abuse,
> To show their Parts, will scold and rail,
> Like Porters o'er a Pot of Ale.

In the light of these meticulous distinctions and the insistence on the social and ethical advantages of the art, one recognizes all the more sharply those occasions when it is bungled or simply set aside. When Swift was at Market Hill, 'the neighbouring Ladies', according to Faulkner, 'were no great Understanders of Raillery'. Market Hill was a rural estate and the implication here may be that country cousins are no match for urban wit. But it seems that these same urbanities frequently wore thin. Sheridan angered Delany and, in the end, lost Swift's friendship with raw remarks that were ill-taken. Participants in the humorous verse 'battles', which were a feature of the intellectual circle, not infrequently lost their tempers; Swift was at various times culprit, victim and peacemaker. His own poem 'The Journal' (1721), describing in terms of affectionate raillery the household activities of the Rochforts and their guests, led to his being criticized for abusing the hospitality of friends. The casualty rate could, admittedly, have been higher; but the point would seem to be that, notwithstanding the precise distinctions between fine raillery and coarse insult, mistakes were frequently made, even by such skilled practitioners. It may seem that infringements occurred through the necessity to turn in small tight circles of mutual exacerbation and the obligation to demonstrate superior skill. Despite accidents this high-strung technique had its value in a close community, permitting the reiteration of such real or imagined values as accuracy, tact and feeling. It offered a way of beating the bounds between the permissible and the unspeakable, of driving out the drones and instructing the rest. In a letter to William Pulteney of 7 March 1736–7 Swift acclaimed his recipient's 'more than an old Roman Spirit' which had been 'the constant Subject of Discourse and of Praise among the whole few of what unprostituted people have remain[ed] among us'. Swift had a tendency to count heads. 'Few' and 'unprostituted' is what the drama of unity required, in the teeth of human limitation, for the sake of the *manes*. He wrote to the Earl of Oxford, 3 July 1714:

> For in your publick Capacity you have often angred me to the Heart, but, as a private man, never once. So that if I only looktd towards my self I could wish you a private Man to morrow.

In one sense this is telling praise; in another, it is a courteous and restrained lament for the failure to achieve unity of Being. In one important respect, the mutual regard of two 'unprostituted' people is everything; but at the same time, given the ideal integrity of public life, it is not enough. Perhaps this is the value for Swift, emotionally and philosophically, of the defeated man and it may be the truly creative thing that he received from the ideas and example of Sir William Temple. Defeat restores Unity of Being, if only hermetically and in isolation. Although critics quite properly stress the manifest differences between the early Pindarics and the mature verse, it is possible to detect a thread of continuity, a line of development starting with the Pindaric celebration of the defeated man (Temple; Sancroft), continuing through the 1716 poem to Oxford, 'How blest is he, who for his Country dies', and culminating in the comic exorcisms of Swift's own defeat. His political and ecclesiastical embarrassment, his 'exile', were factors to which he personally refused to succumb, but which, as a poet, he provoked into a series of difficult encounters. The situation of 'An Apology to the Lady Carteret' (1725) and 'Verses on the Death of Dr. Swift' (1731) is defeat, either by bodily humiliation or the trivia of daily encounters or the triumph of philistine life, but wit converts the necessary failure into moral and rhetorical victory. The prime significance of Swift's 'sin of wit' is that it challenges and reverses in terms of metaphor the world's routine of power and, within safe parentheses, considers all alternatives including anarchy.

Swift's attitude towards the anarchic was significantly ambiguous. In principle he abhorred all its aspects, from lexical and grammatical to social and political; pragmatically he played along with it to some extent; poetically he reacted to it with a kindling of creative delight. While, in the main, anarchy was a mob-attribute, it is open to suggestion that Swift also recognized a Jonsonian sense of disorder, an imbalance of humours in those who governed policy or money. Viewed in this light, great bad men like Marlborough or Walpole were a threat to the body-politic. Although Swift, in common with other contemporaries, offered analyses of raillery and although a subsequent process of scholarly abstraction has tended to give such discussions an apparent universality in Augustan literary and social debate, his work is noteworthy for its so-called 'ungenerous' and 'inexcusable' attacks on Marlborough and somewhat lesser figures like Baron Cutts and Archbishop Hort, poems in which he abandons the salon of raillery for the pillory of invective. There seems little to distinguish the tone of 'The Character of Sir Robert Walpole' or the anti-Marlborough piece 'On the Death of a Late Famous General' from the kind of procedure that Swift, in 'To Mr. Delany', purported to despise. It should be stressed that in considering the poems themselves, when one notes the distinction between raillery and invective one is describing the difference between two equal forces rather than between superior and inferior kinds. If 'Verses on the Death of Dr. Swift' is the apotheosis of raillery it is equally apparent that 'The Legion Club' (1736) is Swift's masterpiece of invective:

> KEEPER, shew me where to fix
> On the Puppy Pair of *Dicks*;
> By their lanthorn Jaws and Leathern,
> You might swear they both are Brethren:
> *Dick Fitz-Baker, Dick* the Player,
> Old Acquaintance, are you there?
> Dear Companions hug and kiss,
> Toast *old Glorious* in your Piss.
> Tye them Keeper in a Tether,
> Let them stare and stink together;
> Both are apt to be unruly,
> Lash them daily, lash them duly . . .

Invective is a touchy subject. Sir Harold Williams, to whose researches all students of Swift are indebted, writes of 'The Legion Club' as an 'uncontrolled outburst', but it is difficult to see quite what he means. In these lines Swift describes the vicious as being also, in a sense, helpless. 'Stare' perfectly expresses fixated energy. 'Old Acquaintance' seems equally well judged: the two 'Companions', Richard Tighe and Richard Bettesworth, were well-known opponents of Swift, participants in lengthy feuds. In the poem he makes them 'old' in the sense that the devil is 'old' Nick; that is, their sinful madness is inveterate. This is the fusion of familiar and formal in a word. Or consider:

> Bless us, *Morgan*! Art thou there Man?
> Bless mine Eyes! Art thou the Chairman?
> Chairman to yon damn'd Committee!
> Yet I look on thee with Pity.

Here an 'impossible' rhyme (there Man/Chairman) toys with the 'impossibility' of finding a decent man like Morgan in such a place, while its thumping obviousness simultaneously confirms his presence. The magisterial tone quite transcends the real source of Swift's outrage, a pot-and-kettle dispute between the Irish clergy and the landowner-dominated Irish Parliament over the question of pasturage-tithes. It is poetically convincing and technically invulnerable; not an 'uncontrolled outburst' but in places a deft simulation of one. Because it is so convincing it could even be said to react upon its source. While admitting the parochial nature of the original feud, one is prepared to accept that universal principles of human conduct – justice, dignity and right dealing – are here involved and that Swift's protest is uttered on behalf of common honesty and freedom.

*   *   *

Supposedly 'uncontrolled' outbursts also affect that most sensitive area of Swiftian research, the so-called 'unprintable' poems. There is a line of defence on these which laudably aims to explode the pathological fallacy but requires careful qualification. It has been said that Swift 'is hardly more scatological than others of his contemporaries' and Professor Ehrenpreis has drawn attention to possible parallels in Dryden's version of Juvenal's 'Tenth Satire', in a burlesque (dated 1702) of L'Estrange's *Quevedo's Visions* and in *Le Diable Boiteux* of Le Sage. Parts of Smollett's *Humphry Clinker* and *Adventures of an Atom* also constitute admissible evidence. But when all has been allowed Swift still remains more comprehensively and concentratedly scatological than his English contemporaries. One cannot seriously compare such squibs as the anonymous 'On a F[ar]t Let in the House of Commons' with 'Cassinus and Peter', or 'Strephon and Chloe' with the mild and modish pornography of Prior:

> At last, I wish, said She, my Dear –
> (And whisper'd something in his Ear.)

'Whisper'd something' is truly symptomatic of the mode and perhaps helps to explain by contrast the nature of Swift's verse, which cuts through that barrier of shame and coquetry where it is only too easy to excite a snigger with gestures of mock reticence. The best of Swift's scatological poems can therefore be called 'harrowing' in the true sense of that word. It is open to argument that the best are those which are most susceptible to accusations, on the part of hostile critics, of violent morbid obsession. The range of these poems is extremely varied and Swift at his worst is quite capable of polite innuendo, superior to Prior in verbal *brio* but hardly superior in ethos.

It may be proper, as a preliminary step, to establish the robustness of Swift's scatological humour:

> My Lord, on Fire amidst the Dames,
> F[art]s like a Laurel in the Flames.

Here it subsists in the comedy of 'on Fire' and 'like a Laurel', in the suggestion of genial heat and sparkling Olympian success. It occurs in later work such as 'Strephon and Chloe' (1731):

> He found her, while the Scent increas'd,
> As *mortal* as himself at least.
> But, soon with like Occasions prest,
> He boldly sent his Hand in quest,
> (Inspir'd with Courage from his Bride,)
> To reach the Pot on t'other Side.
> And as he fill'd the reeking Vase,
> Let fly a Rouzer in her Face.

'He boldly sent his Hand in quest' is the language of lyrical pornography applied to an unlyrical situation. The catharsis of the episode has been well described by Maurice Johnson who speaks of 'the surprise of the line in which that word [Rouzer] appears, startling the reader to laughter or at least to exclamation'. It would be difficult to find a word that blends the outrageous and the festive more effectively than this. Swift is capable of outrage at the world of spontaneous reflexes but he is equally offended by the false notions of 'divinity' previously entertained by Strephon about women and by Chloe about men; hence the real importance of the perception in '*mortal*'. In a basic sense these anarchic explosions are more real than the sublimities of attenuated fancy but they are still grotesque, Swift suggests, when contrasted with the proper decencies and restraints of life:

On Sense and Wit your Passion found,
By Decency cemented round . . .

But the very fact that these basic functions do have this element of truth in
the situation, and come with a mixture of unexpectedness and inevitability,
produces something of the festive energy of farce. However deliberately the
retrenching moralist may stand at guard (e.g. the routine phraseology of the
concluding lines) the poetic imagination still cherishes the creatures of its
invention.

It would be a mistake to enter a plea for Swift's defence merely on the
evidence of 'healthy laughter', which is something of an evasive term and
fails to cover the dominant characteristics of Swift's major scatological
work. It could be said that the embarrassments of Strephon and Chloe are
trivial and susceptible to robust treatment but that 'The Lady's Dressing
Room' (1730) and 'A Beautiful Young Nymph Going to Bed' (1731) reveal
an appalled, and appalling, obsession with filth and disease; that we are far
from the delighted surprise of 'Rouzer' in lines like these:

> The Bason takes whatever comes
> The Scrapings of her Teeth and Gums,
> A nasty Compound of all Hues,
> For here she spits, and here she spues.

Although this may be basically comparable to passages in Dryden and Le
Sage, its emotional and verbal concentration is undeniably unique. So far as
accusations of simple 'bad taste' are concerned, there is no great difficulty in
showing that, in terms of the eighteenth-century conditions of life, it would
be virtually impossible to exceed plain reality. Objections on this ground
alone are like the tasteful reservations of Swift's aristocratic friends, such as
the Earl of Orrery, who saw, but didn't enjoy what they saw. On the other
hand it is true that the language of such poems may seem excessive when
compared with Swift's equally unflinching work in *A Modest Proposal*. There
he can cover the small circle of a mean intelligence (the 'persona') with the
ample radius of the intellect itself and make his tacit reservations tense with
humane implication. In 'A Beautiful Young Nymph' it may seem that the
superior intelligence can assert itself only by extravagant gestures of
revulsion. Notwithstanding these objections, a modern critic who appeals to
the supposed imperative of compassion is gesturing towards little more
than a modern version of pastoral. To fret at its absence from Swift is really
to miss the point:

79

> The Nymph, tho' in this mangled Plight,
> Must ev'ry Morn her Limbs unite.
> But how shall I describe her Arts
> To recollect the scatter'd Parts?
> Or shew the Anguish, Toil, and Pain,
> Of gath'ring up herself again?
> The bashful Muse will never bear
> In such a Scene to interfere.
> *Corinna* in the Morning dizen'd,
> Who sees, will spew; who smells, be poison'd.

The perfect dryness of 'recollect', the charged portentousness of 'dizen'd', inviting and awaiting the sharp crack of the final rhyme-word, have complete control over the plangencies of 'Anguish, Toil, and Pain'. If one argues for compassion here it has to be admitted that it exists principally for the eye of the beholder; and this is not one of the forms acceptable today, when pity is all for the object. One would also observe that if Swift's view of Corinna is scarcely charitable neither is it unfeeling.

Professor Ehrenpreis has convincingly answered a certain kind of objection to the emotional intensity of some of Swift's savagely satirical verse by remarking:

> the complainants' case would be best proved if Swift were *not* intense on such subjects.

The accuracy of this as a general ethical observation can be sustained, even though the immediate defence-plea for Swift is damaged, by citing 'A Panegyrick on the Dean, in the Person of a Lady in the North' (1730). This poem does not seem to be among those generally selected for attack, though it ought to be. Composed as a contribution to the mirth of the Achesons at Market Hill, it is cast in the form of an address of commendation to the Dean by Lady Acheson herself. One of its main themes is the erection of two latrines on the estate and the consequent partial reformation of manners:

> Yet, some Devotion still remains
> Among our harmless Northern Swains;
> Whose Off'rings plac't in golden Ranks,
> Adorn our chrystal River's Banks:
> Nor seldom grace the flow'ry downs,
> With spiral Tops, and Copple-Crowns:
> Or gilding in a sunny Morn
> The humble branches of a Thorn.

One could, in theory, defend 'A Panegyrick' as being an elaborate but proper complaint about the recalcitrance of human behaviour: the 'swains', having been offered hygiene, still prefer their old casual filth. The poem, however, eludes this kind of defence. If it is set against 'The Lady's Dressing Room' or 'A Beautiful Young Nymph', the intensity of the sensuous attack in those poems appears as a valid human reaction; one is convinced that the virtuosity of description is necessary to contain and express the range of Swift's feelings. 'A Panegyrick', on the other hand, is Swift's only scatological poem that seems in any sense coprophilous. It is, significantly enough, the one which most nearly approaches the conditions of *salon* verse. The poem's tonality suggests that Swift is writing, not out of fascinated disgust or angry contempt, but under the obligation to amuse: it is the very coolness of the verbal draughtsmanship, the detailing of the faecal coils, that is so chilling. The fact that Swift parodies the namby-pamby of fashionable pastoral is, I think, beside the point so far as the larger issues are concerned. 'A Description of a City Shower' also parodies pastoral but this fact neither explains nor limits that poem's total effect. In 'A Panegyrick' any plea of parody is really an alibi for the indulgence of a taste that is itself more dubious than 'straight' pastoral. It is a pity that terms like 'perfectly calculated' have become, out of context, mere laudatory commonplaces in criticisms of Swift. As a generalization this could apply to a great deal that he wrote; but between the perfect calculation of 'A Description of a City Shower', 'Verses on the Death of Dr. Swift' or 'The Legion Club' and the perfect calculation of 'A Panegyrick' there stretches a wide terrain of ethical and aesthetic distinction.

It is possible that 'perfectly calculated' could, in some instances, be better expressed as 'predetermined' or even 'academic'. The academic Swift is a significant figure, notably in *A Proposal for Correcting, Improving and Ascertaining the English Tongue*, an attempt to predetermine the future shape of the language in accordance with his overriding political convictions and apprehensions. Swift's linguistic attitude is a kind of Tory stoicism, a rather simpler form of resistance than that of the poems. Its limitations are perhaps best described by Dr Johnson, in the Preface to his *Dictionary*. Johnson remarks on the futility of trying to secure language from corruption and decay and of imagining that one has the 'power to change sublunary nature, and clear the world at once from folly, vanity, and affectation'. One recalls Yeats's saying that Swift 'foresaw' Democracy as 'the ruin to come'. If this is so, it only intensifies the creative paradox of his poetry whose energy seems at times to emerge from the destructive element itself. In his own copy of Dr Gibbs's paraphrase of the Psalms Swift scribbled unflattering comments alongside examples of slovenly rhyming. Of 'more-pow'r' he commented: 'Pronounce this like my Lady's woman'. Yet in Swift's own poetry this lexical and grammatical arrogance is transfigured, as in such a work as 'The Humble Petition of Frances Harris' (1701). Mrs Harris was one of Lady Berkeley's gentlewomen:

> Yes, says the *Steward*, I remember when I was at my Lady
> Shrewsbury's,
> Such a thing as this happen'd, just about the time of
> Goosberries . . .
>
> The *Devil* take me, said she, (blessing her self,) if I ever
> saw't!
> So she roar'd like a *Bedlam*, as tho' I had call'd her all
> to naught . . .

Some years later, in 1718, Swift wrote a further poem, 'Mary the Cook-Maid's Letter to Dr. Sheridan', along similar lines of character and idiom. A number of his polemical pieces adopt, or affect, the form and phrasing of the popular street-ballads of the day: 'Peace and Dunkirk: Being an Excellent New Song upon the Surrender of Dunkirk' (1712); 'An Elegy upon the Much-Lamented Death of Mr. Demar, the Famous Rich Man' (1720); 'An Excellent New Song Upon His Grace Our Good Lord Archbishop of Dublin' (1724); 'Clever Tom Clinch Going to be Hang'd' (1726); 'The Yahoo's Overthrow; or the Kevan Bayl's New Ballad . . . to the tune of "Derry down"' (1734). The Bagford, Pepys and Roxburghe Collections contain possible precedents and analogues, such as the anonymous 'South Sea Ballad' of 1720, which could be compared and contrasted with Swift's poem, 'The Bubble', of the same year. If we refer to the confrontation of two distinct kinds of poetic tradition and method, a popular and an aristocratic, and if we relate Swift's Pindaric 'aberration' to the kind of sublimity invoked by Sir William Temple in his *Essay on Heroick Virtue* (1690) there will be a temptation to claim that a timely encounter with popular colloquial verse 'redeemed' Swift as a poet. But there is no simple and obvious way in which this could be affirmed. Some of Swift's poems may have achieved immediate popular success, but one still has reservations about calling him a 'popular' poet; he did not so much use as demonstrate the colloquial; the very kind of accuracy he achieved was the result of a certain aloofness. He was able to fix his perspectives:

> And so say I told you so, and you may go tell my Master;
> what care I?
> And I don't care who knows it, 'tis all one to *Mary*.
> Every body knows, that I love to tell Truth and shame the
> Devil,

> I am but a poor Servant, but I think Gentle folks should
> be civil . . .

If one is conscious, throughout, of the dualistic nature of Swift's genius and achievement, it is not inappropriate that so much of his energy should have been expended upon, and re-charged from, the dualistic nature of Irish life and politics, or that one should be able to find both the cold disdain and the fervent identification of the poems coexisting in the style of *The Drapier's Letters*, whose effect on the Irish people is happily evoked by Lecky's general encomium on Swift:

> He braced their energies; he breathed into them something of his own lofty and defiant spirit; he made them sensible at once of the wrongs they endured, of the rights they might claim, and of the forces they possessed.

Given the current English political attitude, to be in Ireland, the 'depending kingdom', as a member of the so-called governing class was to be in a 'situation' of considerable difficulty. Swift polemically rejected the situation as a principle in the fourth *Drapier's Letter* but encountered it daily as a fact. His sensitive reaction to this situation, both personal and national, resulted in a release of creative energy which could not have been produced by the application of principle alone. Swift had little sympathy with Shaftesbury; but the crowning paradox is that his own poetry is one of the most powerful expressions in eighteenth-century English literature, prior to Blake, of that kind of resistance which the Whig philosopher so eloquently described:

> And thus the natural free Spirits of ingenious Men, if imprisoned and controul'd, will find out other ways of Motion to relieve themselves in their *Constraint* . . . 'Tis the persecuting Spirit has rais'd the *bantering* one . . . The greater the Weight is, the bitterer will be the Satir[e]. The higher the Slavery, the more exquisite the Buffoonery.

SIX

Redeeming the Time

Miss Iris Murdoch, in her cogent and illuminating *Sartre*, makes an uncharacteristically elusive assertion when she claims that:

> The smaller, expanding world of the nineteenth century, where the disruptive forces were not only dispossessed and weak, but incoherent, disunited, and speechless, could think itself a single world wherein rational communication on every topic was a possibility.

Faced with such a statement by so manifestly intelligent a writer one looks nervously for a hidden language-game: e.g., the century could think itself a single world, could cherish the hypothesis; but probably it did not; or, if it did, perhaps it deceived itself. Some major questions are begged, not only as to the area of possibility but also as to the limits of metaphor in 'speechless'. Henry Adams, it is true, remarked that 'Beyond a doubt, silence is best' but Adams, in his several substantial volumes, was the patentee of one of the most obliquely cautious of modern styles; and in any case he was dispossessed only in an acutely specialized sense of the term. Those who were dispossessed in the more mundane demotic fashion, deprived of rightful franchise, income and leisure, tended to be vociferous rather than speechless. Dr E. P. Thompson, a scholar by no means devoid of sympathy for the protagonists of proletarian struggle, has scrupulously drawn attention to 'that demagogic element, inevitable in a popular movement excluded from power or hope of power, which encouraged the wholly unconstructive rhetoric of denunciation'. R. H. Tawney, equally sympathetic towards the principles of democratic progress, himself wrote of 'the orgy of mob oratory in which Chartism finally collapsed'. The common factor present in Iris Murdoch's generalization and in the statements of Tawney and Thompson is far from insignificant. Even so, anything less like a 'single world' than the nineteenth century it would be hard to discover; and without moving outside the circumference of Iris Murdoch's model arena one encounters the violent clashing of contradictions. It is at least open to suggestion that the epoch was marked by a drastic breaking of tempo and by an equally severe disturbance of the supposedly normative patterns of speech.

It should not be thought, however, that in referring to the possibility of a 'language-game' one is being lightly dismissive, or underrating those implicit and explicit tactics whereby a class or faction might contrive to project itself as 'the world'. Charles L. Stevenson has reminded us that 'Ethical definitions involve a wedding of descriptive and emotive meaning, and accordingly have a frequent use in redirecting and intensifying attitudes'. The confines of a determined world 'give' so as not to give; tropes are predestined to free election; the larger determinism allows for the smaller voluntarism. Theodore Redpath, describing Locke's presentation of the institution of 'lawful government', shows how that writer is appealing to his readers

> by the use of a compelling *image* – the image of a body which cannot move except in the direction in which 'the greater force' carries it. The image is clearly drawn from dynamics, and would have had a special appeal in the age of Newton.

Granted the operation of a post-Newtonian dynamics of appeal, the by-products of entropy may also be expected: among them the trivia of decayed good taste and the 'demagogic . . . rhetoric of denunciation':

> Here I stand the Factory King, declared King by the most contemptible enemies of the Cause. Yorkshire is mine! (*Cheers.*) Lancashire is mine! (*Cheers.*) Scotland is mine! (*Cheers.*) All Christendom is mine! (*Cheers.*) WE WILL HAVE THE BILL . . .

These 'feverish words' are Richard Oastler's. Mr Cecil Driver, in his scholarly and sympathetic *Tory Radical: The Life of Richard Oastler*, attributes such utterance to:

> the striking change of disposition that had come over Oastler in the months since he had first discovered his platform powers . . . that at least a part of his personality found a dangerous delight in the thunders of the multitude it was now too obvious to doubt.

Our indebtedness to Mr Driver suffers no diminution if we fail to endorse his general suggestion. 'Disposition' and 'personality' are insecure bases upon which to build speculation, whereas the nature of a man's occupation, the range of his expectations and the limits of his security might well be influential in forming the rhythms and cadences of his speech. Richard

Oastler, like the father of George Eliot, was a land-agent; he was an agrarian ultra-Tory and a devout Anglican; he was also resident in the West Riding of Yorkshire at the time of its greatest social upheaval. To allude to the range of his expectations and the limits of his security is neither to impugn the disinterested courage of his actions nor to bring the discussion round in a full circle to speculation about individual disposition and personality. England has been described, in a congested metaphor, as 'the classic soil of ... transformation' from a largely agrarian to an urban proletarian society, and whether one chooses to let the phrase stand, with all its thronging connotations, or to isolate the three terms 'classic', 'soil' and 'transformation', Engels's words are not inappropriate to the fortunes and misfortunes of nineteenth-century English styles. Oastler's words are decidedly unfortunate. When allowance has been made for the effect of partisan enthusiasm or prejudice upon the style of contemporary reporting, Oastler remains caught in the involuntary comedy of his voluntary stance. Phrases redolent of heroic Protestantism ('here I stand', 'enemies of the Cause') and presumably chosen for that reason, are degraded into a mummer's rant. The 'classic' oration of a tribune of the people suffers a grotesque 'transformation', 'soiled' by the outflow of a fractured tradition. A rooted man, faced by the uprooted, by a floating population of the new proletariat, is unable to prevent his own words floating. Their hectic nature has possibly less to do with any personal fever than with the contagion of circumstance. Elsewhere in radical utterance a more circumstantial manner of address is itself an acknowledgement of determinism, as in the declaration of the Nottingham branch of the United Committee of Framework-Knitters, *c.*1812:

> It is true that Government has interfered in the regulation of wages in times long since gone by; but the writings of Dr. Adam Smith have altered the opinion, of the polished part of society, on this subject. Therefore, to attempt to advance wages by parliamentary influence, would be as absurd as an attempt to regulate the winds.

By 'circumstantial' one means that the authors of this passage try to account for circumstances and also that they 'go about' their direct task with language that is itself enclosed, 'fenced round' by idioms of the society which determines the realities of the situation. They are writing up, and writing up to, the subject; and writing up involves the acceptance of terms that have come down. The canting locution 'polished' had been thoroughly drubbed by Johnson as early as the middle years of the eighteenth century. Johnson was familiar with the brutality and drudgery underlying the polished life but he was also his own master and capable of exerting mastery.

His sardonic stoicism resurfaces, in the working-class Nottingham of 1812, as an intelligent but etiolated fatalism. In his view 'the polished part of society' seemed vain and foolish; the framework-knitters of 1812 know that the polish is that of adamant.

An enquiry into the nature of rhythm must first attempt to account for the inertial drag of speech. Language gravitates and exerts a gravitational pull. Oastler is not 'compelling' but compelled. Social locutions which to others might be scarcely more than half-comic irritants impose upon the Nottingham framework-knitters a force as shiftless as that of nature itself. In Wordsworth's 'Ode: Intimations of Immortality . . . ', published in 1807, the line

> Heavy as frost, and deep almost as life!

is a weighed acknowledgement of custom's pressure; stanza eight is allowed to settle onto this line. However, the poet immediately breaks continuity, thrusts against the arrangement, the settlement, with a fresh time-signature

> O joy! that in our embers
> Is something that doth live . . .

It has been pointed out that in this poem 'the prevailing rhythm is merely iambic' and the Ode has been further described as 'broken-backed'. Saintsbury may be technically correct; but Wordsworth's strategy of combining a pause with a change of time-signature within the 'merely iambic' prevailing rhythm overrides both the propriety and the pressure. It could be suggested, in response to C. C. Clarke's criticism, that the Ode is indeed broken but that the break, far from being an injury sustained, is a resistance proclaimed. If language is more than a vehicle for the transmission of axioms and concepts, rhythm is correspondingly more than a physiological motor. It is capable of registering, mimetically, deep shocks of recognition. This quality in Wordsworth's Ode was finely perceived by Gerard Manley Hopkins and explained in his letter of 23 October 1886 to Canon Dixon:

> There have been in all history a few, a very few men, whom common repute, even where it did not trust them, has treated as having had something happen to them that does not happen to other men, as having *seen something*, whatever that really was . . . human nature in these men saw something, got a shock . . . in Wordsworth when he wrote that ode human nature got another of those shocks . . . his insight was at its very deepest, and hence to my mind the extreme value of the poem . . . The rhymes are so musically interlaced, the rhythms so happily succeed (surely it is a magical change 'O joy that in our embers') . . .

Hopkins's praise is to my mind in no way extravagant and his remarks constitute a major contribution to the study of rhythm. Wordsworth's 'insight was at its very deepest' and he 'saw' rhythmically. To do justice to the quality of his seeing one must refer again to Richard Oastler and to the Nottingham framework-knitters. Writers on linguistics employ a term 'stress-pitch-juncture'. '"Juncture" is that particular configuration of pause and pitch characteristics by which the voice connects linguistic units to each other or to silence.' In this case one requires a modified term, 'stress-pitch-disjuncture'. Oastler, instead of being able to subsume the satirical attacks of his opponents, is made, through his own verbosity, an accessory to their mockery. The Nottingham men are made to mouth the vulgarisms of their betters and to repine half-heartedly for 'times long since gone by'. Energy and perception have been driven apart. Perception has turned to fatalism, energy has flung into frenzy. If Wordsworth has indeed 'seen something' he has seen, or foreseen, the developing life-crisis of the nineteenth century. In the Ode the shock to be suffered by Oastler and the Nottingham men, among many others, is redeemed by the silence between stanzas eight and nine and by the immediate, abrupt surge with which the 'joy' of nine's opening lines resists, pulls away from, the gravitational field of the closing lines of stanza eight. Wordsworth transfigures a fractured world. As Hopkins truly said 'surely it is a magical change "O joy that in our embers"'. This is one of the rare contexts in which the debased word 'magical' regains some of its pristine power. Yet although 'magical' is allowable, 'change' is the key-word and should take precedence. To show why this is so, some comparison is necessary.

George Eliot is a writer with a fine sense of traditional rhythmic life, as is shown by chapter 18, 'Church', of *Adam Bede*:

> But Adam's thoughts of Hetty did not deafen him to the service; they rather blended with all the other deep feelings for which the church service was a channel to him this afternoon, as a certain consciousness of our entire past and our imagined future blends itself with all our moments of keen sensibility. And to Adam the church service was the best channel he could have found for his mingled regret, yearning, and resignation; its interchange of beseeching cries for help, with outbursts of faith and praise – its recurrent responses and the familiar rhythm of its collects, seemed to speak for him as no other form of worship could have done . . .

Here, 'recurrent responses' is a wide-ranging term. The responses are to be understood both as recurring within the limited time-span of a particular Anglican evensong, following the established pattern of the rubric, and as recurring over an implied and indefinite number of years, Sunday by

Sunday, season by season. It is by such means that 'channels' are created; by the joint working of abrasion and continuity. 'Responses' is the correct term for the established form of congregational participation in the liturgy. At the same time, over and below this literal meaning, the word connotes the continuity of human response in general to an ancient process of parochial and national life. The collects of the Anglican Church are composed of liturgical prose; they could properly be said to possess rhythm, though not metre. Here again, however, 'familiar rhythm' is both liturgical and extra-liturgical, telling of a rhythm of social duties, rites, ties and obligations from which an individual severs himself or herself at great cost and peril, but implying also the natural sequences of stresses and slacks in the thoughts and acts of a representative human being. In George Eliot's last book, *Theophrastus Such* (1879), there is an essay called 'Looking Backward' in which the fictitious narrator is the son of a country parson in the Midlands. In this essay George Eliot writes significantly of 'the speech of the land-scape'. It is as though, at the end of her creative span, the author could compress into a phrase of five words the essence of paragraphs and chapters of earlier work such as *Adam Bede* (1859). The ethics are right – whether or not one happens to share them is immaterial – she is happy with the theme and the theme is happy with her. It is of course 'organic' and half-feudal, imbued with 'affectionate joy in our native landscape, which is one deep root of our national life and language'. Such a style can unite the strength of the 'deep root' with an evocation of the concomitant torpor of 'fat central England'. To praise the quintessential mastery of this late essay is not to abandon, nor even to qualify, one's praise of chapter 18 of *Adam Bede*, which could fairly be discussed under the general heading of 'poetics'.

It is instructive and saddening to set alongside these passages of masterly conflation such an example of blatant, disingenuous compounding as is the 1868 pamphlet *Address to Working Men by Felix Holt*:

> But I come back to this: that, in our old society, there are old institutions, and among them the various distinctions and inherited advantages of classes, which have shaped themselves along with all the wonderful slow-growing system of things made up of our laws, our commerce, and our stores of all sorts, whether in material objects, such as buildings and machinery, or in knowledge, such as scientific thought and professional skill . . . After the Reform Bill of 1832 I was in an election riot, which showed me clearly, on a small scale, what public disorder must always be; and I have never forgotten that the riot was brought about chiefly by the agency of dishonest men who professed to be on the people's side.

The source of one's objection must be clearly defined: it is in an area where misunderstanding is all too easy. An early critic of the piece protested that

'Felix Holt the Radical is rather Felix Holt the Conservative; he is not even a Tory-Democrat'. It would not be within our area of debate to attack the pamphlet on *a priori* political grounds. If George Eliot is a conservative here she is equally so in the beautiful essay 'Looking Backward'. The falsity of the pamphlet lies in its rhythmic gerrymandering and not in its basic code of beliefs. George Eliot has denied us the cross-rhythms and counterpointings which ought, for the sake of proper strategy and of good faith, to be part of the structure of such writing. In short, she has excluded the antiphonal voice of the heckler. Felix's argument is fair enough but it ought to be fairly heckled, as for instance: 'in our old society, there are old institutions . . . which have shaped themselves' (antiphonal voice of heckler: 'Shaped themselves? how? as naturally and as easily as leaves on the tree?'). Or: 'I have never forgotten that the riot was brought about chiefly by the agency of dishonest men' (heckler: 'Name three.'). George Eliot has denied us 'the drama of reason'.

The phrase is Coleridge's. On 28 January 1810 he wrote to Thomas Poole. He noted that he had been studying *The Spectator* 'with increasing pleasure & admiration' but immediately qualified this praise with the suggestion that Addison's paper had 'innocently contributed to the general taste for unconnected writing'. 'Innocently' seems in every sense a judicious emphasis. Coleridge's suggestion seems to be that a style which, around the year 1710, issued from and upheld a genuinely humane sensibility had, by 1810, been run down into the 'fixities and definites' of a mere 'law of association', into the inert 'general taste' and cliché-ridden fancy that served to gloss over the barbarous prejudice of 'the polished part of society'. In the same letter of 1810 Coleridge defends his own style by anticipating likely objections:

> Of Parentheses I may be too fond – and will be on my guard in this respect –. But I am certain that no work of empassioned & eloquent reasoning ever did or could subsist without them – They are the *drama* of Reason – & present the thought growing, instead of a mere Hortus siccus.

For one whose 'thought' has been popularly presented as fetching 'wide circuits' and as coming 'to no visible end' Coleridge maintained a striking continuity and consistency in his meditation of many years upon the drama of reason. L. C. Knights has remarked that the pamphlets in support of Peel's reformist factory bill of 1818 are distinguished by the 'union of specific information, strict logic, and a basic humanity', and this encomium may be both endorsed and extended by reference to a wide range of distinct but correlative statements throughout Coleridge's work. Of crucial significance is his desiderated 'moral *copula*' which would, he believed, 'take from history its accidentality and from science its fatalism'. His sense of the moral copula, though not exclusively grammatical, was attuned to the minute particulars of grammar and etymology:

> For if words are not THINGS, they are LIVING POWERS, by which the things of most importance to mankind are actuated, combined, and humanized.

The words 'actuated, combined, and humanized' do more than offer a linguistic endorsement of L. C. Knights's ethical criteria; they also take the strain against 'the general taste for unconnected writing':

> On some future occasion more especially demanding such disquisition, I shall attempt to prove the close connection between veracity and habits of mental accuracy; the beneficial after-effects of verbal precision in the preclusion of fanaticism, which masters the feelings more especially by indistinct watch-words . . .

At this point the question might well be raised as to what precisely we have in mind when approving Coleridge's disapproval of 'unconnected writing'. It would be a fair question and one hopes that it could be fairly answered. As Coleridge's reserved praise, or laudatory reservations, about Addison's style serve to indicate, the issue cannot be reduced to a simplistic quarrel between epochs. On the contrary, some lessons on the right conduct of the moral copula might be drawn from an author whose style Coleridge considered 'detestable', from Gibbon's use of the conjunction 'but':

> The architecture and government of Turin presented the same aspect of tame and tiresome uniformity; but the court was regulated with decent and splendid economy.

> By fencing with so skilful a master, I acquired some dexterity in the use of my philosophic weapons; but I was still the slave of education and prejudice.

Of the first of these examples it might be said that it is a bland modulation of the 'trimming' style. If that is so, it must be conceded that the trimming is so openly and palpably effected as to preclude objection. Exception cannot be taken, because exemption has not been claimed. The style which pronounces judgment is open to receive it. In the second example Gibbon achieves a plain style which, in a basic and necessary sense, establishes empirical connectives. As such it is to be fairly distinguished from persuasion by means of 'tendentious equivocation' or 'compelling . . . image' or by means of 'wit', as in Chesterfield:

91

[Chesterfield is] best satisfied when he has reduced his idea to the *atom* of prose, the 'detached thought,' which is the natural medium of wit.

As Coleridge's 'copula' could be said to have been anticipated in Swift or Johnson or Gibbon, so his own criteria are frequently reminiscent of Augustan ideals. His definition of prose as 'words in their best order', the emphasis upon 'good sense' and 'very plain language' have obvious precedents. It is the very strength with which he endorses such qualities that underlines the merit of his readiness to risk convolution and incommunicability in the effort to draw men *'thro'* . . . words into the power of reading Books in general'. It was apparent, even later, in Victorian England, that the eighteenth-century patrimony of speech remained a source of mixed blessings and embarrassments:

> [Newman's] language is that of the ordinary educated persons of his day. [His terms] are drawn from the tradition of British empiricism. They had become so much a part of educated speech that it is a nice question whether they ought to be considered technical terms at all.

These comments are intended to apply only to Newman's Anglican period. That his later style was a more complex procedure is suggested by Walter E. Houghton. Referring to Newman's method of argument in two sentences of the *Apologia* Houghton remarks:

> The structure here is so intricate and involved, the meaning so hard to follow at first reading, that one is tempted not to praise the style but to charge it with unnecessary awkwardness and obscurity . . . And yet, in return for that sacrifice (and it is a deliberate sacrifice: he could, of course, write with perfect clarity of form when he chose), Newman's structure has caught and projected the very sense of wavering, of being pulled back and forth and forth and back, which he was undergoing.

Houghton's analysis is well glossed by John Beer who notes that, at this specific point, Newman's writing 'diverges from the normal pattern of Victorian prose by the undertow of reluctance in its arguments'. This is excellently said, provided that we do not lose sight of the fact that the words are immediately applicable to two sentences and that Houghton's own footnotes refer us to, at most, another half-dozen brief passages of comparable pattern. To point to the sparsity of examples is not to imply a reproach. Newman's proper economy is admirable and one may see it anticipated by Coleridge in the letter of 1810:

I shall endeavor to pitch my note to the Idea of a common well-educated thoughtful man, of ordinary talents; and the exceptions to this rule shall not form more than one fifth of the work—. If with all this it will not do, well! And *well* it will be, in its noblest sense: for *I* shall have done my best.

Houghton's 'deliberate sacrifice' which has particular sacramental relevance in the case of Newman is by no means irrelevant to a consideration of Coleridge, whose passages of self-criticism, which claim precedent and authority from the example of Hooker, strike the reader as being diagnostically resolute rather than symptomatically irresolute:

> If I may dare once more adopt the words of Hooker, 'they, unto whom we shall seem tedious, are in no wise injured by us, because it is in their own hands to spare that labor, which they are not willing to endure'.
>
> The metaphysical disquisition at the end of the first volume of the *Biographia Literaria* is unformed and immature; it contains the fragments of the truth, but it is not fully thought out. It is wonderful to myself to think how infinitely more profound my views now are, and yet how much clearer they are withal. The circle is completing; the idea is coming round to, and to be, the common sense.

However much irony and reservation is directed towards Coleridge's real and hypothetical neuroses as the source of his convolutions and hesitancies, the rectitude of his decision seems unassailable. He surely foresaw the obligation to enact the drama of reason within the texture of one's own work, since nothing else would serve. His parentheses are antiphons of vital challenge.

The value of the antiphonal style was perceived by Matthew Arnold in 'The Function of Criticism at the Present Time':

> Mr Roebuck will have a poor opinion of an adversary who replies to his defiant songs of triumph only by murmuring under his breath, *Wragg is in custody*; but in no other way will these songs of triumph be induced gradually to moderate themselves . . .

Arnold's sentence is a little scenario and *Wragg is in custody* is abruptly expressive within the scenario. 'Defiant' is an epithet ironically accorded Mr Roebuck, whose 'triumph' is then defied by Arnold 'under his breath'. It is a brief comic masterstroke in counterpoint. Its effectiveness exposes all the more drastically the failure of an adjacent passage:

has any one reflected what a touch of grossness in our race, what an original shortcoming in the more delicate spiritual perceptions, is shown by the natural growth amongst us of such hideous names, – Higginbottom, Stiggins, Bugg! . . . by the Ilissus there was no Wragg, poor thing!

The indignation has glided aside into something much less. One can plot the swerve, the graph of descent. 'Wragg is in custody' appals Arnold, and rightly, because it speaks with the voice of the beadle, the complacent harshness of the penal code lopping off 'the superfluous Christian name', a process endorsed by the jubilant tribunes of the vox populi. However, the name Wragg itself strikes Arnold's sensitive ear as horridly vulgar; the critic who has warned against catch-words is caught by a word and, in an unguarded moment, righteous anger and unrighteous taste become compounded. The indignation of a just and compassionate man is degraded into a whinny of petty revulsion. Is she 'poor thing' because of her hideous life or because of her 'hideous' name? In 1836 a factory inspector had discovered a Rochdale weaver 'passionately fond of ancient history' who had named his daughters in accordance with his passion. 'But only think', wrote the inspector, 'what a word was added to each, a word which the poor weaver could neither change nor modify; Barraclough – Pandora Barraclough!'

The issue would seem to be between two forms of sacrifice: sacrifice *of* or sacrifice *to*. The first involves making a burnt offering of a powerful and decent desire, the desire to be immediately understood by 'a common well-educated thoughtful man, of ordinary talents'. Its structure is a recognition and a resistance; it is parenthetical, antiphonal, it turns upon itself; its most consistent practitioners, in the nineteenth century, are Coleridge, in the prose works, and Hopkins; isolated, significant examples are to be found in Newman and T. H. Green. Its text might be taken equally from Coleridge, Green or Hopkins. Hopkins wrote to Bridges on 6 November 1887:

> Plainly if it is possible to express a sub[t]le and recondite thought on a subtle and recondite subject in a subtle and recondite way and with great felicity and perfection, in the end, something must be sacrificed, with so trying a task, in the process, and this may be the being at once, nay perhaps even the being without explanation at all, intelligible.

Hopkins wrote elsewhere to Bridges 'it is true this Victorian English is a bad business' and his remark may be interpreted in two interrelated ways. His own syntax, in the letter of 6 November 1887, is a bad business; it is, to apply one of his own terms, 'jaded'. If we were to enquire why, apart from the commonly invoked private neurasthenic reason, Hopkins's style should be so near the end of its tether, we might well find the answer in a different form of bad business. In 1832, John Stuart Mill observed that 'a certain laxity in the use of language must be borne with, if a writer makes himself understood'. He added the important rider that 'to understand a writer who uses the same words as a vehicle for different ideas, requires a vigorous effort of co-operation on the part of the reader'. If Mill may be regarded as speaking here for a dualistic mode of acceptance, Hopkins may be considered as standing for the minority mode of non-acceptance. The 'general taste' of which Coleridge wrote is, as he knew, no innocent datum but something vicious, even if 'innocent'; to an extent the innocence compounds the vice. This much was evident as early as Burney's review of the 1798 *Lyrical Ballads*, and Burney's tonality may have been haunting Wordsworth's mind when he complained, in the 1802 additions to the Preface, of

> men who speak of what they do not understand; who talk of Poetry as of a matter of amusement and idle pleasure; who will converse with us as gravely about a *taste* for Poetry, as they express it, as if it were a thing as indifferent as a taste for rope-dancing, or Frontiniac or Sherry.

George Eliot was aware of the detrition of general taste. Mrs Transome, in *Felix Holt*, was one of those who, in youth, 'had laughed at the Lyrical Ballads and admired Mr Southey's Thalaba' and her laughter is certainly meant, in retrospect, to ring hollow.

It is the difference, essentially, between vital and inert structures. If these terms appear portentous and suggestive of a 'certain laxity', they may be partly redeemed by close illustration. The 'magical change' between stanzas eight and nine of the 'Immortality' Ode is vital; Oastler's speech and Arnold's 'has anyone reflected' are inert. To these categories one must add a third: where a malign activity is made possible by the very inertia of general taste:

> It is, in fact, the constant aim and tendency of every improvement in machinery to supersede human labour altogether, or to diminish its cost, by substituting the industry of women and children for that of men; or that of ordinary labourers, for trained artisans.

The *OED* defines 'or' as 'a particle co-ordinating two (or more) words, phrases, or clauses, between which there is an alternative.' Scarcely ever can a 'particle' have been employed with such brutal power as in Dr Ure's suave persuasion. Suavity of this nature casts its silvery filaments around those virtues which nineteenth-century artisans were permitted to cherish. George Howell visited the Great Exhibition of 1851:

95

I cannot express my feelings as I entered that vast palace of iron and as I glanced around the multifarious and magnificent collection of the products of the world there represented. All dreams of fairy land were eclipsed in a moment.

If one takes issue with this it is because one can look beyond Howell's 'dreams of fairy land' to Dr Ure's hive of industry.

If Fancy deals, as Coleridge says, with 'fixities and definites', then Howell's delight at the Crystal Palace is fanciful, not imaginative. One's protest is directed not so much against as on behalf of Howell's dream. It is the blatant coexistence in nineteenth-century England of his 'I cannot express' and Dr Ure's 'or' that distresses. Howell's words may be juxtaposed with a radical statement of some twenty years earlier:

How strange that machinery should have an inverted and continually diverging effect upon society, rendering the condition of those attendant upon it worse and worse while others are reaping its amazing productiveness in pernicious luxury.

The two passages are in some respects strikingly complementary: 'amazing productiveness' correlates with 'multifarious and magnificent' but far more noteworthy than any similarity is the antithetical and divergent nature of what is argued. The Huddersfield men are perceptive ('diverging' is their word, and it is right) but they are also overborne, and know it. They are angry yet half-numbed by the monstrous nature of the world. It is in the syntax: 'How strange' introduces a sequence that ought, in justice, to be far-fetched but is in fact the glacial drift of their lives. George Howell, in contrast, is not overborne but rather borne up on the warmth of his astonishment and delight. Historians will properly point to a relative easing of conditions in the 1850s, as contrasted with the 1830s. When that much has been conceded we are left with Howell's ballooning platitudes, further symptoms of a diremption between perception and utterance, energy and effect. It is as though an iron wedge had been driven between the two passages.

There is some justification here for Coleridge's stress upon the 'moral *copula*' to 'take from history its accidentality, and from science its fatalism'. It might perhaps be added that the significance of Coleridge's distinction between primary and secondary imagination, particularly when read in the light of later pronouncements in *Table Talk*, is that the first represents an ideal democratic birthright, a light that ought to light every person coming into the world. In the event, the majority is deprived of this birthright in exchange for a mess of euphoric trivia and, if half-aware of its loss, is instructed to look for freedom in an isolated and competitive search for possessions and opportunity. Therefore the secondary imagination, the formal creative faculty, must awaken the minds of men to their lost heritage, not of possession but of perception. Within certain contexts, such as that of Ure's *Philosophy of Manufactures*, even such virtues as 'rigorous self-improvement' might be fanciful rather than imaginative. F. M. Leventhal, describing the influence of Methodist teaching on the young George Howell, remarks that:

> the prescription of obedience and industry concealed an underlying ambiguity. Without denying that moral indiscipline could lead to social rebelliousness, it was equally clear that moral rectitude encouraged self-reliance. The logical outcome of education and perseverance was ambition for self-betterment, not a humble acceptance of inferior status.

It could be said, however, that 'ambition', like 'humble acceptance', is more than a matter of belief or attitude; it is a matter of what is available or allowed. Even so, F. M. Leventhal's choice of terms is helpful. It is at least open to suggestion that each of the terms or concepts, 'obedience', 'industry', 'indiscipline', 'logical outcome', adumbrates its own distinctive rhythm, or rhythmic disjuncture, and that an 'underlying ambiguity' governs the nature of rhythm as much as it does the nature of ethics.

Hopkins wrote that he employed sprung rhythm as being 'nearest to the rhythm of prose, that is the native and natural rhythm of speech . . .' In the context which this present discussion has attempted to establish it could be argued that the citation of a uniquely 'native and natural rhythm of speech' is itself highly selective, even ideological. This is not said in reproof. A study of the underlying ambiguity of nineteenth-century society enhances rather than detracts from one's respect for Hopkins's achievement. If asked to explain in more detail the nature of this alleged ambiguity, one would turn for support to authoritative discussions of 'the commodity status of time' and of 'mechanical time versus organic time'. The opportunity to refer to such extensive, complex and subtle arguments permits one's own suggestion to be made much more simplistically. Crudely stated, the difference is between being 'in' stride and 'out of' stride. The 'magical change' in the 'Immortality' Ode is perhaps the greatest moment in nineteenth-century English poetry; but in choosing this term one is suggesting restriction as well as potency. The recognition and the strategy to match the recognition – the cessation of 'stride', the moment of disjuncture, the picking up of fresh 'stride' – were of their very nature inimitable; they were of, and for, that moment. It could be said, however, that in his choice of themes and methods, Hopkins is attempting a correlative pattern. The achievement of sprung rhythm is its being 'out of stride' if judged by the standards of common (or running) rhythm, while remaining 'in stride' if considered as procession, as pointed liturgical chant or as shanty. In 'Harry Ploughman' the man is in stride, his craft requires it; and the poem itself, in its rhythm and 'burden lines', is the model of a work song. In the companion-piece, 'Tom's Garland', the dispossessed are thrown out of work and out of stride and the piece is, both discursively and rhythmically, perhaps the harshest, most crabbed, of all Hopkins's poems. It is as though the poet is implying that, because the men cannot work, therefore the poem itself cannot. Hopkins's persistent sense of being 'jaded' may, as has already been conceded, have neurasthenic causes; but this concession does not diminish the respect that is due to him for encompassing in his rhythm not only 'the achieve of, the mastery of the thing' but also 'the jading and jar of the cart'. 'Tom's Garland' is a failure, but it fails to some purpose; it is a test to breaking point of the sustaining power of language.

In arguing for Hopkins's vital perception of the underlying ambiguities in nineteenth-century speech rhythms one comes to recognize the central importance of two phrases in particular. They are 'abrúpt sélf', in 'Henry Purcell', and '(my God!) my God', in '[Carrion Comfort]'. For Hopkins man is revealed in his intense selfhood and in his most frightful splintering. In the contemporary *Times* reports of the 1875 wreck of the *Deutschland* Hopkins could have read of a desperate man hacking at his wrist with a penknife in the hope of a comparatively painless death by bleeding, of another hanging himself behind the wheelhouse, of a last message placed in a bottle, even as he read of a gaunt nun calling out 'O Christ, come quickly!' and 'My God, my God, make haste, make haste' till the end came. Dom David Knowles quotes the anonymous author of *The Cloud of Unknowing* on the power of short prayer:

> A man or a woman, affrighted by any sudden chance of fire, or of a man's death, or whatever else it be, suddenly in the height of his spirit he is driven in haste and in need to cry or to pray for help. Yea, how? Surely not in many words, nor yet in one word of two syllables. And why is that? Because he thinketh it over long tarrying, for to declare the need and the work of his spirit. And therefore he bursteth up hideously with a great spirit, and cryeth but one little word of one syllable: such as is this word FIRE or this word OUT.

Dame Helen Gardner in her essay 'Walter Hilton and the Mystical Tradition in England' also glosses the essential meaning of the author of the *Cloud*:

the mind must be emptied of all thought, save that which is contained in the short words 'God' and 'Love', and by this unknowing alone can God be known and loved, not God in His goodness or in His mercy, but '[the] nakid beyng of him'.

If we remove the phrase 'nakid beyng' from its immediate and proper theological context and apply it somewhat impressionistically, we may consider that it has a strong connotative impact, and, further, that the connotations aid our study of Hopkins. 'Aid our study' is tentative; one could not legitimately go beyond it. The following remarks proceed by analogy, not by evidence. Man, as well as God, could be a 'nakid beyng': man destitute; essential man. Wordsworth uses 'naked' in this double sense.

In a letter to his mother, written from Stonyhurst, 2 March 1871, Hopkins devotes half the available space to a demonstration of the characteristic Lancashire intonation 'Ay!' The letter is designed to entertain and divert the recipient but the overt intention does not conceal the close attention paid to the phonology and physiology of the utterance. Hopkins describes the conversation of two gardeners:

> What the one says the other assents to by the roots and upwards from the level of the sea. He makes a kind of Etna of assent, without effort but with a long fervent breathing out of all the breath there is in him. The word runs through the whole scale of the vowels beginning broad in the barrel of the waist and ending fine on the drop of the lip. For this reason I believe it is a natural sign of agreement and not conventional . . . It is always intoned.

Intonation can refer to the 'manner of utterance of the tones of the voice in speaking' as in 'that unfortunate intonation of Aberdeenshire'. 'An "intonation pattern" is the amalgam of features of stress, pitch, and juncture which occur as part of a spoken phrase.' It is also a technical term in church music: 'The opening phrase of a plain-song melody, preceding the reciting-note, and usually sung either by the priest alone, or by one or a few of the choristers.' The range of the two words 'scale' and 'intoned' in Hopkins's letter is potentially that of the two words 'responses' and 'rhythm' in chapter 18 of *Adam Bede.*

Here again, however, the differences are as significant as the resemblances. A rhythm of 'interchange' was available to George Eliot as it was not to Hopkins. To say this is not to ignore either her self-excommunication from the Church of England or the appalling social ostracism endured by Lewes and herself. In 1859, the year of *Adam Bede,* she wrote in a letter of 'a sympathy . . . that predominates over all argumentative tendencies. I have not returned to dogmatic Christianity . . . but I see in it the highest expression of the religious sentiment that has yet found its place in the history of mankind'. Being able to think in these terms – Hopkins would have considered it a sloppy form of idealism – enabled her to stay imaginatively, if not actually, 'in stride' with Anglican parochial and national life. The power of this Anglican 'rhythm' should not be underrated:

> The medieval cathedrals and churches, the rich ceremonies that surround the monarchy, the historic titles of Canterbury and York, the social organiz-ation of the country parishes, the traditional culture of Oxford and Cambridge, the liturgy composed in the heyday of English prose style – all these are the property of the Church of England . . .

These are not the words of an Anglican apologist; they are those of the late Evelyn Waugh, explaining the initial sense of loss sometimes experienced by English converts to Rome. Hopkins wrote to his father, 16 October 1866:

> I am surprised you shd. say fancy and aesthetic tastes have led me to my present state of mind: these wd. be better satisfied in the Church of England, for bad taste is always meeting one in the accessories of Catholicism.

Michael Trappes-Lomax had pointed out that in the mid 1830s, at about the time of Pugin's conversion, the Mass, 'except for the private chapels of rare wealthy Catholics and a few other places, was said in garrets, in tawdry assembly-rooms, in lofts over stables'. It would be wrong to suggest that the tawdriness was the sole prerogative of Catholics. Before his submission to Rome Pugin had been disgusted by the damage done to English churches either by neglect or by the 'folly and arrogance' of misguided restorers. He had been equally horrified by the manners of some of the Anglican clergy: he complained that 'the Rev.—goes to perform the service in *top boots* and *white cord breeches*' and that the son of a bishop had 'lost £7,000 at the last Lincoln races'. I do not think that such evidence alters the basic fact that Anglicanism, however debased and abused, could offer a 'rhythm' of responses and that the English convert to Rome, however much he might gain, nonetheless suffered an abruption of this familiar rhythm. George Eliot, in *Adam Bede*, as in the late essay 'Looking Backward', establishes a pattern of inherited living, in which the interchange of expectation and limitation constitutes the private drama. In Hopkins virtually the reverse is true: he begins with 'nakid beyng' and proceeds to build up a 'scale'. Father Walter J. Ong, S J, has commented:

> If Hopkins' claims for his rhythm are acceptable, he must have been, consciously or unconsciously, hearing it everywhere . . . the rhythms of a

language are already rooted when the poet arrives, and the real question to be answered concerning Hopkins' sprung rhythm is, What was this thing he was discovering all around him?

One of the aims of this present discussion has been to question, albeit tentatively and inadequately, the premiss of 'rooted' speech-rhythms. This may seem an astonishing suggestion in the light of the phrase 'assents to by the roots' in the Stonyhurst letter of 2 March 1871. My opinion, which is of course open to challenge, is that the essential word is the reiterated 'assents . . . assent'. To assent by the roots is to become an entire embodiment of *assent*. What seems to have delighted Hopkins was the simple coherence of spirit, voice and body: a humorous corollary to his most intense concern. When D. P. McGuire states that Hopkins 'comprehended . . . the characteristic features of the oncoming world: the still increasing tempo . . . of modern life' his terms merit serious consideration. One answer to Father Ong's question 'what was this thing [Hopkins] was discovering all around him?' could be 'increasing tempo'; another answer could be 'the ambivalent power of the "short words"'. The short words are neither rooted nor uprooted, graced nor ungraced; they may go with the 'tempo' or they may be made to react against it. They are the most elemental material, and they are the abrupt selving of prayer: 'We lash with the best or worst/Word last!'

The task of the Catholic poem, at least as practised by Hopkins, could be seen as corresponding to the task of the tall nun who 'christens her wild-worst Best'. The poet, who made sensitive reference in his letters to the gifts and sacrifice of the Elizabethan Jesuits Campion and Southwell and who must have known also of the staunchness and death-agonies of the Carthusians martyred at Tyburn in 1535, must have sensed also how the morphemes of demotic speech, the irreducible 'Ay!' of subsistence, could be reconciled both with the mystic discipline of 'short prayer', with the significance of the Corpus Christi procession, and with the sustained melody of Gregorian chant. In the death-cry of Prior Houghton there is for us a dreadful mingling of physical agony and willing oblation, of helplessness and horror struggling to transmute themselves into a voluntary sacrifice. To one with Hopkins's theological and poetic discipline the cries could be more meaningfully (it would be offensive to say more easily) assimilated into the liturgy and emblems of martyrdom:

The copy of the *Summa* which Campion was using . . . survives at Manresa College, Roehampton; it is annotated in his own hand and opposite an argument on baptism by blood occurs the single *mot prophète et radieux*, '*Martyrium*'.

The 'short word', 'mot prophète et radieux' ('ah! bright wings'; 'ah my dear'), tackles the brutality, buffoonery and mere obtuseness of English and transfigures it. 'Wragg, poor thing' or 'WE WILL HAVE THE BILL' are there in '"Some find me a sword; some / The flange and the rail; flame, / Fang, or flood", goes Death on drum', 'They fought with God's cold-', 'And wears man's smudge and shares man's smell'; and they are answered by 'I did say yes' and by 'Christ, King, Head'. What we have termed the ambivalent power of the short words is most eloquently realized in the final line of 'Carrion Comfort': '(my God!) my God.' The abrupted experiences once more commune with each other: the expletive of a potentially filthy bare forked animal ('I wretch', 'carrion') and the bare word of faith.

'Extremes meet': the appropriate term is Hopkins's own. He is writing of his own art, and of Whitman:

> Extremes meet . . . this savagery of [Whitman's] art, this rhythm in its last ruggedness and decomposition into common prose, comes near the last elaboration of mine ['The Leaden Echo and the Golden Echo'].

This may not be good Whitman criticism but it is good Hopkins criticism. In suggesting that our 'rhythm' enacts our life-process, or that life-process involves rhythm ('you cannot eat your cake and keep it: he eats his off-hand, I keep mine. It makes a very great difference'), he verges upon a symbolism of society and of history as pertinent and as vulnerable to challenge as those of Henry Adams and W. B. Yeats. Unlike them he had no need to extend his perception into a formula. It was his gain but in some respects, possibly, our loss. The tension between 'off-hand' and 'keep', and between 'eat' and 'keep' is an essential tension, particularly when the partaking of the elements at Mass is borne in mind. There is a consuming which is part of the process of 'organic' dissolution, and there is an absorption which is not. In a letter written some time afterwards, Hopkins noted:

> The only good and truly beautiful recitative is that of plain chant; . . . It is a natural development of the speaking, reading, or declaiming voice.

'Decomposition' (explicit) is poised against 'composition' (implicit); we break down into 'common prose', we build, we scale up from, common prose. Taking the two letters together one can see that the pattern is too fluid to be 'rooted'; at the same time one recognizes that 'fluid' is suggestively approximate to those processes of dissolution which the theologian Hopkins rejected. One senses again that the letter to his mother on the morpheme of Lancashire speech and the letter to Bridges on the Corpus Christi procession may be more significant to a study of his poetry than has perhaps been realized. 'Ay!' could be simply inclusive of passion and belief; the Corpus Christi procession was sacramentally inclusive of passion and belief. Highly popular as well as richly liturgical, it did not spill over into the demotic but drew the demotic in. Possibly the best description of Hopkins's poetic method would be his own 'recurb and . . . recovery'.

All this could legitimately be termed 'paternalistic', 'conservative', even 'reactionary'. Such descriptions would not be value-judgments on the poetry itself. One word which, in my opinion, ought not to be applied is the word 'decadent', yet it figures prominently in a distinguished and widely-known essay on Hopkins:

> Hopkins wrote in a decadent age, and if he is its greatest poet, he may be so because he cultivates his hysteria and pushes his sickness to the limit. Certainly he displays, along with the frantic ingenuity, another decadent symptom more easily recognized, the refinement and manipulation of sensuous appetite . . . Much of his work, in criticism and poetry alike, is concerned with restoring to a jaded palate the capacity for enjoyment.

Donald Davie's critical diction is here richly allusive rather than 'pure'. The phrase 'cultivates his hysteria' is an allusion to the entry in Baudelaire's Journal, 'I have cultivated my hysteria with delight and terror'. If one wishes to imply decadent 'sensuous appetite', would it not be more accurate to speak of a 'sated' rather than of a 'jaded' palate? Hopkins referred to himself as 'jaded' – exhausted by overwork and by nervous stress. To use this word and to suggest that it is precisely synonymous with satiety seems to me just a little tendentious.

That Hopkins wrote in a decadent age is unarguable; but to see him as quintessential of that condition is to fail to comprehend what decadence truly is. In contrast to Coleridge and Hopkins, it was 'decadent' of Mill to concede 'a certain laxity' for the sake of ease of communication: acquiescence requires quiescence. Ure's *Philosophy of Manufactures* is a degrading work. It was 'decadent' of Arnold to stoop to the indulgence of his more delicate spiritual perceptions: with all his intelligence and compassion he could not stop himself falling for the sneer, and such 'falling' is 'decadence'. It was decadent of George Eliot to write two separate essays, 'Address to Working Men' and 'Debasing the Moral Currency', the first of which betrays the priorities implied in the second, with no sense of contact or coherence between them. To fall for these things or to conspire so that others fall for them is decadence. Against all this Hopkins's poetry established a dogged resistance. Both ethically and rhythmically, his vocation was to redeem the time.

SEVEN

'Perplexed Persistence':
The Exemplary Failure of T. H. Green

T. H. Green could be said to have provided an epigraph for his own work when he wrote of Butler that 'his value as an ethical writer is due to the same cause which makes his speculation perplexed and self-contradictory'. Praise of Green is not infrequently accompanied by major reservation, and when it is unqualified it is of dubious value, as for instance, '[he] sent us out once again on the high pilgrimage towards Ideal Truth', where the symptomatic 'high', as in 'high idealism', has a fulsomeness which Green for the most part succeeded in avoiding in his own work. 'He would not have liked high language such as this to be applied to himself' wrote Nettleship, of the obituary tributes; though, in a significant concession, he added 'but it is true'. In so far as Nettleship's comments demonstrate a confrontation between Green's respect for concreteness and his immediate admirers' zeal for abstraction, they may be taken as a marginal indication of the central issues in Green's metaphysical and ethical debate. 'Abstract the many relations from the one thing, and there is nothing' he wrote in *Prolegomena to Ethics*, as elsewhere he argued that Burke revealed 'true philosophic insight' when he opposed the 'attempt to abstract the individual from ... the relations embodied in habitudes and institutions which make him what he is'. The phrase 'concrete experience' is to be found in Green before it is found in I. A. Richards.

That Nettleship should find the concession necessary is indicative of the strained relations between intention and reception which prompted some of Green's best writing and speaking but which also rendered him liable to those blurrings and evasions which detract from the power of his ethical style. He criticized Butler for being 'content to leave the moral nature a cross of unreconciled principles' while, as a corollary, he argued that 'Man reads back into himself, so to speak, the distinctions which have issued from him, and which he finds in language' and that, in this 'retranslation', he 'changes the fluidity which belongs to them in language, where they represent ever-shifting attitudes of thought and perpetually cross each other, for the fixedness of separate things'. Green shares here a prevalent ethical emphasis of his time, the recognition that while we are 'uncon-

104

ditionally bound', 'necessarily belonging to such a world', being so bound is not necessarily the same as being in a fix and is most certainly not the same as being a fixer. In his dual application of the word 'cross', once as noun and once as verb, in two consecutive paragraphs, Green finds words for an essentially Kantian crux. The nature of the world is such as we are constrained to recognize, the ineluctable fact, but to be content with the rich discrepancies which this offers is none the less dangerous and is sometimes treacherous. If, in making this emphasis, Green is equal to the strength of such contemporary ethical art as George Eliot's, he also shows symptoms of contemporary weakness and rigidity. Both aspects may be examined in relation to an observation by Coleridge which stands as a paradigm for some of the most significant Romantic and post-Romantic debate on the supposed 'formal engagement' between literature and society. Coleridge takes a passage from one of Donne's sermons which, he argues, 'sways & oscillates', in its use of the word 'blood', between a 'spiritual interpretation' and a physical sense: 'Yea, it is most affecting', Coleridge writes, 'to see the struggles of so great a mind to preserve its inborn fealty to the Reason under the servitude to an accepted article of *Belief*...' There are times when it seems that Green himself is marked by some such discrepancy between 'fealty' and 'servitude'. This is not to argue that he was uniquely deficient but rather to suggest how the common struggles of the time took a particular form in his work. In *Prolegomena* he claims that we apprehend knowledge as we 'apprehend the import of a sentence ... In reading the sentence we see the words successively [but] throughout that succession there must be present continuously the consciousness that the sentence has a meaning as a whole'. He further contends that we need to make 'constant reference to the expression of that experience which is embodied, so to speak, in the habitual phraseology of men, in literature, and in the institutions of family and political life'. It could be argued that Coleridge's objection is applicable here and that Green's 'so to speak' and 'phraseology of men' oscillate between a literal and a metaphoric sense. Pragmatic obedience to the structure of the sentence, in order to apprehend a part of that sentence, is not the same as adherence to a metaphor of grammar, the 'habitual phraseology' of men, unless it is decreed that it shall seem so. 'So to speak', in this context, enables him to accept the decree without necessarily promulgating it.

Though the cast of his ethical thought was opposed to the currently dominant forms of adherence and coercion, Green himself too often wrote in those 'vague but insistent' terms which Whitehead has justly categorized as 'social symbolism': 'Thus the "Treatise of Human Nature" and the "Critic of Pure Reason", taken together, form the real bridge between the old world of philosophy and the new.' Although, as Green noted elsewhere, 'every explanation ... involves a metaphor', 'real bridge' here pre-empts its own verification. It lacks the resonance of those 'hypnotizing' terms to which Whitehead draws attention, but its tacit premiss is still the 'response of action to symbol' which is at the centre of that strategy. And this strategy is itself crucial in certain forms of nineteenth-century Liberal bridge-building. In the words of J. S. Mill: 'We hold that these two sorts of men [Benthamites and Coleridgeans], who seem to be, and believe themselves to be, enemies, are in reality allies'. Although 'we hold that' and 'in reality' are set obliquely to each other, Mill's phrasing is designed to conceal the angularity of his proposition. It is 'in reality' which exerts the greater persuasive rhetoric and effectively overrides the proviso in 'we hold'. Later in the century than either Mill or Green, the economist Alfred Marshall wrote that economic studies 'call for and develop the faculty of sympathy, and especially that rare sympathy which enables people to put themselves in the place, not only of their comrades, but also of other classes'. Here again it could be said that such phrasing is so much a part of innocuous common parlance that to object is a sign only of hermetic irritability. And it is true that we have no good reason to dissent from MacKinnon's recognition of 'the language of ethics' as a 'language of prescription, exhortation, persuasion, moral tradition' with 'its own texture and laws'. But of even greater significance is his caveat that we must 'recognise in the authority [of the moral law] which . . . unconditionally constrains us the voice of our own nature as rational beings, and not the brusque, sheerly unintelligible dictate of a despot'. Green's 'real bridge', Mill's 'in reality' and Marshall's 'in the place of' seem attempts at an illicit bridging of 'the chasm which the Kantian analysis of judgment left between subject and object'. And there is, in this form of illicit persuasiveness, a hint of the despotic. In such instances we are not so much 'unconditionally constrained' as constrained to be conditioned; it is less a matter of language living in usage than of usage perching and hatching parasitically upon the surface of language. It would be unjust to argue that such devices are the product of cynicism. At the same time it would be naïve to contend that sincerity confers absolution. Towards the end of his life Green sincerely anticipated a time 'when every Oxford citizen will have open to him at least the precious companionship of the best books in his own language'. 'At least' betrays a desperation that the overt optimism will not concede. 'Best' by whose judgment, or according to what criteria? Does it mean of incontestable formal excellence, even though shocking to some susceptibilities (e.g. *Les Fleurs du Mal*), or does it mean incontestably good-hearted, like Mrs Humphry Ward's *Robert Elsmere*, which was dedicated to Green's memory? There can be little doubt that his allegiance was to that 'compound of optimism and humanism', a typically mid-Victorian variety of what William James called 'the religion of healthy-mindedness'. Green was a Liberal who objected to the more materialistic forms in which ideas of Liberal progress took shape during his lifetime. He deplored, in *Prolegomena*, a state of society in which so many were 'left to sink or swim in the stream of unrelenting competition, in which we admit that the weaker has not a chance'. In the words of H. J. Laski, 'Green and his followers emphasized not the individual over against the process of government, but the individual in the significant totality of his relations with it'.

There is a case to be made for the suggestion that one of the major discoveries of modern criticism has been the method of transferring 'significant totality of relations' to a contextual plane and of conferring a consequent distinction upon those authors or individual works which fulfil most completely that kind of expectation: Keats, in the Odes, George Eliot in *Middlemarch*, or the poems of Green's sometime pupil Gerard Manley Hopkins. If that is indeed the case, then modern criticism is ratifying an inclination which has been present in ethical writing since at least the time of Arnold. When Arnold praised a paragraph in Burke's 'Thoughts on French Affairs' as 'one of the finest things in English Literature' he reversed significantly an order of precedence. Burke's concern was not with 'literature' in that hypostatically pure form. In Arnold's statement the 'literature' is central, the politics a catalyst in the creation of the integral substance. This is not to suggest that Arnold was unconcerned with the politics but rather that he saw literature as containing politics within a sphere of more precisely adjusted anxieties. Green is both better and worse than Arnold. On the whole he refuses Arnold's trim formulas and offers a closer reading of the patterns of relationship. But even if it is now widely customary to read a text in the way that Laski suggests that Green's ideal state could be 'read', Green's own work is not amenable to this form of approach. It could be left an open question whether on this issue he rebukes modern criticism or modern criticism rebukes him, since there is a proper reluctance to bestow on any one critical position the power of seemingly absolute arbitration. Even so there is in Green a confused thinness analogous to, though not identical with, the superficiality of secular evangelism, from Mill to J. M. Keynes, which encouraged what the latter called 'the civilising arts of life' while largely disregarding their stubborn textures. In 1889 the theologian P. T. Forsyth commented that 'the best preaching analyses its text, and even discusses points of its grammar' but what sometimes passed for the best preaching in the nineteenth century was in some respects deficient in that kind of analysis. Since Green and Sidgwick were celebrated for the fervency of their utterance, their own failure to meet the terms of Forsyth's criterion merits some examination. In a letter written in 1868 Sidgwick said: 'Oh, how I sympathise with Kant! with his passionate yearning for synthesis and condemned by his reason to criticism . . .' With its heavily plangent tone this sentence speaks more for the moodiness of mid-Victorian intellectualism than for the mood of Kant. The manner in which it yearns over the dichotomy tells us a good deal about the characteristic gestures of Sidgwick and of Green also. It is open to suggestion that in a deep sense they soften Kant's rigour while, at a superficial level, they accommodate his rigorous tone. The difference between Kant and the Victorian students of Kant in England may well turn on a difference of emphasis concerning that which 'points beyond the data'. In Kant this is a 'common element' existing as 'a logical presupposition, a purely formal implication'; in Sidgwick and Green that which points beyond the data is more often a pious wish. Pious wishes are of course wholly valid, unless they are presented as logical presuppositions and purely formal implications. It is then that they cease to be 'that which points beyond the data' and become the 'vague ultimate reasons' which Whitehead has so precisely described. These slight yet massive differences oblige us to draw yet again upon the Coleridgean distinction between fealty and servitude and to say that Sidgwick and Green, while appreciating the distinction in principle, confused its obligations in practice.

It must be said again that to discuss the shortcomings of Green and Sidgwick is to be caught up with ' "*Bona fide* perplexity", as having its origin really in intellectual difficulties, not in any selfish interest'. And in this sense the virtues of Green in particular are difficult to extricate from his shortcomings. At his best he is sufficiently acute as a moral analyst to encourage the suggestion that his self-thwartings and bafflings are acts of homage and abnegation to the Wordsworthian principle. Richter speaks of a 'Wordsworthian sentiment' to be found in him and suggests that the poet was one of the influences disposing Green to 'a pantheistic conception of God as manifest in nature as a spiritual principle'. Green described the 'Ode to Duty' as 'the high-water-mark of modern poetry'. It could be said, however, that there is something more dourly fundamental than 'sentiment', something more stubbornly empirical than even the most high-minded spiritual principle, in Green's Wordsworthian strain. In his essay 'Popular Philosophy in its Relation to Life' (1868) Green proves himself to be a critic of insight and cogency. That quality in Wordsworth which evidently holds Green's critical imagination is one which might be described as the capacity to go against one's own apparent drift:

> In England, it was specially Wordsworth who delivered literature from bondage to the philosophy that had naturalised man. This may at first sight seem a paradoxical statement of the relation between one known popularly as the 'poet of nature' and a system which had magnified 'artifice'. It is not so really.

'It is not so really.' Seemingly laconic, to the verge of inelegance, this phrase offers its own rebuke to the afflatus of 'real bridge' and 'best books', gives the rebuff of imaginative scepticism to fanciful idealism. 'It is not so really' because as Green strongly implies, here and elsewhere, the dogmatic spasms of 'I like' and 'I don't like' can claim no sanction from Wordsworth's popularly misconstrued 'spontaneous overflow of powerful feelings'. 'The natural man is the passive man.' Green's observation is closely followed by his friend and editor, A. C. Bradley, in his own essay on Wordsworth, printed in *Oxford Lectures On Poetry* (1909). At the heart of Bradley's vision of Wordsworth is a sense that the poet comprehended a bi-fold authority: both an acceptance of 'the fixed limits of our habitual view' and a dogged experiential working towards what MacKinnon has called 'the voice of our own nature'. Bradley speaks of 'that perplexed persistence, and that helpless reiteration of a question' which he sees as characteristic of Wordsworth's 'Resolution and Independence'. It is evident, however, that Bradley has some reservations about this 'perplexed persistence' and fears that it may verge on the 'ludicrous' in ways that Wordsworth's 'intimation[s] of bound-lessness' do not. We here touch upon a contradiction that affects Green as much as it affects Bradley. Keenly perceptive of the 'data' which form the texture of a work, each still yearns for the 'vague ultimate reasons' which poetry such as Wordsworth's may seem to provide. Each is inclined to treat weakness as strength and strength as weakness and to override his own shrewdest judgments upon the work. Bradley's words are none the less a fine description of a critical perception and a critical method which are in the poem's own structure. It is worth noting that the line 'Perplexed, and longing to be comforted' does not appear in the original 1807 text; we could say that the poem took time to realize the voice of its own nature in this line. The fact remains that Bradley is asking Wordsworth to de-fuse the source of the ludicrous, as being un-ideal, whereas the poet accommodates the possibility of the ludicrous into the situation and the dialogue. The nar-rator's final pious couplet, though morally worthy, is certainly not the intimation of boundlessness that some readers expect; it has been con-sidered anticlimactic and naïve. In a general observation on Wordsworth, Leslie Stephen said that he 'speaks as from inspiration, not as the builder of a logical system . . . When he tried to argue, he got, as he admits with his usual *naïveté*, "endlessly perplexed".' But Wordsworth, as the opening stanzas of 'Resolution and Independence' reveal, perceived the perils of 'inspiration'; and in his poem he built a logical system to embody a meditation upon the true and false natures of that state of being. His creative gift was to transform the helpless reiterations of raw encounter into the 'obstinate questionings' of his meditated art without losing the sense of rawness. And Bradley is admirable in his own perplexed insight into what

Wordsworth had to do: 'Yet with all this, and by dint of all this, we read with bated breath . . .'

Green has something of this positive quality which Stephen idly termed naïveté. Hastings Rashdall, a critic who nevertheless dedicated a major work to Green's memory, stated that:

> The ethical system of Kant (assisted in England by the influence of Butler and his followers) has produced a hopeless confusion between the question whether Morality consists in promoting an end and the question what that end is.

It could be argued that as Wordsworth transformed helpless reiteration into obstinate questioning so Green, working directly from Kant and Butler and indirectly from Wordsworth and Coleridge, transformed what Rashdall calls 'hopeless confusion' into something more closely resembling perplexed persistence. It had the nature of a physical shock:

> I had the privilege of taking a few essays to Mr. Green . . . I went to his home with my work, and he used to sit over the fire, 'tying himself into knots'. He beat out his music with some difficulty, and the music itself was not an ordinary melody.

Richter says that Green was a bad speaker and cites the opinion of one of Green's admirers, Henry Scott Holland, that he was 'cruelly inarticulate'. Such a view is strongly modified by testimony from those who knew him as well and admired him as much as did Holland. Nettleship wrote: 'Few, I think, can have been more successful in avoiding conversational inadvertence, and the saying of things which had better not have been said'. J. H. Muirhead has described Green's Lay Sermon of 1877 as conveying his ideas 'with singular clearness and with a telling application to the intellectual difficulties of the time'. If Green appeared to some as 'cruelly inarticulate' it can only have been through a form of vocational renunciation, an 'almost confounding humility', a decision as personal yet as formal as that of Hopkins to burn his early poems. Another student has recalled:

> I . . . followed his remarkable lectures with enthusiasm and tense strain . . . I can remember that I did not understand a single word as I wrote down the perplexing tangle of phrases furiously and at lightning-speed: then in the quiet of my rooms I brooded over them till light seemed to gleam from the written word.

Among the words that figure prominently in this and the previously quoted student-memoir are 'music', 'perplexing' and 'gleam', three key-words in

110

'Tintern Abbey'. There is a sense in which Green's students seem to have responded to his words as Wordsworth responded to his own primary experience. It is also worth remarking that 'music' is a term which can be exploited both ideally and empirically. It is the 'still, sad music of humanity' and it is the precise detail of articulation, the 'difficult music' of communication. Coleridge commenced the second essay of *The Friend* by stating that 'an author's harp must be tuned in the hearing of those, who are to understand its after harmonies'. The image of the harper is apt. He is simultaneously a hireling without privacy and a master of pitch and cadence. His concern with tuning is attributable both to his own professional conscience and to his servitude to those who can call the tune. Some of Green's own most cogent objections are to 'the charlatanry of common sense', and to the citing of 'agreeable sensations and reflections' as supposed criteria for determining the value of literature. He regarded 'cultivated opinion' as 'confused' as well as being guilty of 'wilfulness' and shallowness. It could be said that the working alternative to the spasmodic 'I like' and 'I don't like' was embodied in his own seemingly 'inarticulate' utterance. As a tutor to Oxford gentlemen he shared some of Mark Pattison's doubts about the examination system and some of his suspicions that it might encourage a specious fluency. In *Prolegomena* he wrote of the scholar's or artist's 'temptation to be showy instead of thorough'. At the same time, as an educational reformer, Green seems to have sensed that he had a duty to reveal the freedom of the word to those who were, in Wordsworth's term, 'shy, and unpractis'd in the strife of phrase'. It is possible to see an inevitable strain and thwarting in this dual situation: discouraging the wrong sort of fluency and self-display at one level while encouraging the right sort of fluency and self-realization at another level.

Graham Hough has suggested that 'the oral channel was probably the one through which most of the Coleridgean influence originally worked' and he has drawn an illuminating analogy between the power and effect of the Highgate table talk and the influence, at once intimate and far-reaching, of Oxford and Cambridge tutorials conducted by such men as Newman, Jowett, Green and Hare, some of whom were Coleridgeans by direct inheritance, some of whom were not. A necessary proviso is that such access and influence testify not only to intellectual calibre – formidable though that undoubtedly was – but also to the intimate prerogatives of a social élite. At the same time such tutors, Green in particular, were fully aware that they addressed the privileged and they drew such an awareness into their forms of moral instruction. In certain ways, therefore, Green may be regarded as a protagonist in the Coleridgean 'drama of reason' and his speech may be understood as an extension of the original author-reader scenario sketched with much grim humour in the pages of *The Friend*.

In saying this one is perhaps in danger of overstating Green's achievement. Chenevix Trench's *On the Study of Words* (1851) is as steeped in the notion of pastoral care as is Green's fragmentary address 'The Word is Nigh Thee' and Trench was far more orthodox than Green in his religious belief. If we consider 'impulse' alone, then Trench's book could be simply described as an attempt to provide 'valuable warnings ... against subtle temptations and sins'. And yet the book he wrote was far more radical than anything by the 'radical' Green. It was Trench who learned from Coleridge, via Emerson, 'how deep an insight into the failings of the human heart lies at the root of many words'. The difference is both slight and deep. One of Green's prime aims was to resist any tendency towards that fatalism which he saw as the dark side of utilitarian hedonism, towards a condition in which 'there is no alternative but to let the world have its way, and my own inclinations have their way'. He could be regarded, however, as an example of how the anti-utilitarian, anti-hedonist, may yet be held in the gravitational field created by those forces. An acute critic of impulse, Green could counter it only in terms of superior impulse: 'It may very well happen that the desire which *affects* a man most strongly is one which he decides on resisting.' In so doing he endorsed his opponents' standards in the act of challenging them. The very terms he chose called the tune. As a sentence from his testimony before the 1877 Oxford University Commission suggests, he equated 'literary skill' with skilful superficiality: 'Success ... naturally falls to the man of most literary skill, who can bring his mind to bear most promptly and neatly on any subject that may be set before him'. There is therefore an air of frustration, of wasted labour, in Green's attempts to get under the skin of verbal acceptance. Nettleship's 'Memoir' puts on record that 'Those who have ever heard it will remember the peculiar smack of his utterance of the word *tilth*'; and '"Swing" was a favourite word with him to describe the movement of native eloquence, and he would express his dissatisfaction with much contemporary English poetry by saying, with a characteristic gesture of the hand, "There is no *swing* in it"'. This is of course a seductive example because we cannot fail to be aware of the use made of the word 'swing' by Green's occasional pupil Hopkins, in the early poem 'Heaven-Haven' ('And out of the swing of the sea'). But freed from these associations we see that Green's words are merely impulsive, verging on the kind of bluster so effectively parodied by Hopkins when he evoked a manner that 'came in with Kingsley and the Broad Church School', 'the air and spirit of a man bouncing up from table with his mouth full of bread and cheese and saying that he meant to stand no blasted nonsense'.

There is one other haunting fragment of unrealized possibility. Nettleship also records that Green:

had a theory in composing . . . that all superfluous words should be extirpated, the fewest and most compressed used: that, if possible, an essay should consist of one indivisible paragraph, the connected expression of a single proposition or a single syllogism.

But it remains an early 'theory'. Here again it is in the work of his pupil Hopkins, a pupil with whom his relations seem to have been fairly edgy, that we find Green's ideals realized: in the 'indivisible . . . connected expression' of 'Spelt from Sibyl's Leaves' or, perhaps more arguably, in the syntax of the final stanza of 'The Wreck of the Deutschland', with its culminating line which, in Elisabeth Schneider's words, is 'locked together in the hook-and-eye grip of the possessive case':

> Our hearts' charity's hearth's fire, our thoughts' chivalry's
>   throng's Lord.

Miss Schneider gives a striking description, although one recalls that when Coleridge referred to 'all the *hooks-and-eyes* of the memory' he apologized for using 'so trivial a metaphor', sensing perhaps that it might suggest something more simply and mechanistically associative than he intended. Hopkins seems rather to be stressing, against the linear process, an enfolding of possessives, in order, as Miss Schneider so rightly says, 'to create the closest unity of all human values in Christ'. I cannot entirely agree with her suggestion that 'the effect is arbitrary and labored'. The method, I would accept, is arbitrary and laboured but the effect is one of hard-won affirmation.

It seems to me that there is some analogy between the method and effect of Hopkins's poem and the method and effect of Green's lectures. Green's method was repeatedly called 'difficult', but the effect of his words, as we have heard, was likened to 'music' or to a 'gleam'. Admittedly, one is dealing here with subjective impressions. Many of Green's lectures have survived and are printed in the *Works*. The series on 'The English Commonwealth' brings out the 'earnestness' and 'exhilaration of energy' which Green found in the period. For him there appears to have been a 'swing' in it. Elsewhere, his lectures now seem heavy with the diffuseness of paraphrase rather than tense with the bafflements of communication. But the subjective evidence cannot be set aside; it exists in its own right:

Though he had great difficulty in expressing himself at that time . . . Everyone saw that there was great substantial value and originality in the work; and the very difficulty of his utterance gave one the feeling that he was working the thing out, and not repeating other people's phrases or ideas . . .

The men in fact took a sort of pride in the difficult process which he went through before he got things clear, as if it were in some way the joint action of us all.

Even if it is conceded that young men, chasing after charisma, will always hunt down what they seek, through these outmoded Victorian locutions there shines an act of recognition which is stronger than the commerce of mere egotisms. In part, though only in part, 'the joint action of us all' is the burden of the concluding lines of 'The Wreck of the Deutschland', the creation of 'the closest unity of all human values in Christ'. Rather than transaction or projection, a Green lecture appears as an act of atonement, in the arena of communication, between the 'unconscious social insolence' of the listener and what Coleridge termed the seeming 'assumption of super-iority' on the part of the speaker.

It has been said that, for Green, 'the characteristic thing about human experience is that it is thinking experience'. It is a token both of his achievement and of his relative failure that so much of the 'exhilaration of energy', which he manifestly aroused, should be conveyed to us in the form of reminiscence. He himself wrote of the 'weakness . . . which belongs to all ideas not actualised, to all forms not filled up' and several reasons could be advanced for Green's naggingly persistent failure to actualize his own ideas. It could be attributed to the 'hopeless confusion' about means and ends that Rashdall saw as characteristic of English paraphrases of Kantian ethics. Or it could be traced to the overriding confusion as to the nature of the 'formal engagement' in public discourse, a debate in which Wordsworth and Coleridge stand as heroic protagonists. Richter detects, throughout Green's ethics, 'signs of the strain produced by his merger of conservative concepts with liberal and even radical values' and he is, I think, correct in this major respect: the reasons for Green's relative failure are not separate and distinct but manifold. Green, as much as Mill or Sidgwick, was writing to be received, and, at the same time, was conducting a running battle with the premisses of current receptivity. Newman wrote that 'where there is no common measure of minds, there is no common measure of arguments'. Of the nineteenth-century 'Liberal' writers it could be said that they sought a 'common measure' to set 'against mere impulse and mere convention alike', but were left too often with immoderate commonplace. Sidgwick's 'law infinitely constraining and yet infinitely flexible' is no 'law' at all, but a rather shifty rationalization of a condition of servitude. The efforts of Mill, Sidgwick and Green are too easily reducible to the terms of Mill's own assumption that it was good 'to enlighten . . . practice by philosophical meditation', a procedure which leaves pragmatism and idealism as two separate floating entities. It may be that Mill, to use Bagehot's chilling phrase, is feeling the minds of his readers 'like a good rider feels the mouth of his horse', but if that is so it is revealed as a self-defeating exercise. There is some irony in the fact that Mill introduces his essay on Coleridge with such a proposal. 'Meditation' is here regarded as being on a par with 'opinions' and 'mental tendencies'. If practice can be so 'enlightened' by meditation, then meditation can also be forbidden access to the pragmatic domain. What Coleridge gives, however, is a constant demonstration that meditation is central to practice, whereas 'opinion', 'tendency' and 'enlightenment' are peripheral and non-resistant. It has been suggested that one of the dominant philosophical tenets against which Coleridge contended was 'the dogmatic assumption of the principle of *dichotomy*'. If, as Richter has claimed, Coleridge was one of the several exemplars upon whom Green based his own 'faith', it could likewise be suggested that this faith is set out in Green's essay on Aristotle. He writes there of the 'unfused antithesis of the "necessary" and the "contingent"' as manifesting the 'Aristotelian dualism [at its] most conspicuous'; and he ponders upon Aristotle's thesis in such a way as to bring out those elements which are susceptible to a belief in interchange rather than separation: 'The account of the form or essence, then, as a "substance dematerialised" may be replaced by an account of it as a "potentiality actualised"'. There remains in Green's discursive faith, however, a fundamental separation which no amount of theoretical application can bridge. This failure may be termed a failure of imagination because there seems to be no connection in Green's mind between his recognition of an 'account of form' on the one hand and his suppositions about the workings of the poetic mind on the other. Admittedly, he is intent on warning the poet against the perils of euphoria and his words might be taken as a parody of the euphoric state. But there cannot be a parody where there has been no perception of the real condition. We receive no hint from this that Green knew what he was talking about: 'As the poet, traversing the world of sense, which he spiritualises by the aid of forms of beauty, finds himself ever at home, yet never in the same place . . .'

The question 'how the moral intelligence gets into poetry' is of course quite distinct from the question 'how morality gets into poetry' or 'how moral is poetry?' If Green so often disappoints us it is because his critical insights are too frequently dissolved, dissipated, without being re-created. But even this is far from being a simple matter of mere personal vacillation. It could be said that a study of Kant, while encouraging an emphasis upon contextual relationships in general, could nevertheless inspire incompatible applications of the idea of context. 'The new students of Kant in England' drew from their study, among other inferences, the suggestion that '[one] could not, for instance, distinguish even particular items unless [one] were also aware of the context or series in which the items appear'; and it is this sense of 'context or series' which, arguably, is of prime importance in Green's pattern of thinking. The thesis that context distinguishes the items, significant though it might be in projecting an idea of reciprocity between the state and the individual, is nevertheless a definition more at the service of empirical application than of creative imagination. At the same time, 'the form by which spatial and temporal data are apprehended as elements in a whole that points beyond the data' was one which made possible an indifference towards, and even a contempt for, context in anything other than a serial, spatial sense:

> Not as to the sensual ear, nor necessarily through the stinted expression of verbal signs, but as a man communes with his own heart, you may speak to God . . . Prayer indeed, if of the right sort, is already incipient action; or, more properly, it is a moral action which has not yet made its outward sign.

It is possible that 'sensual ear' is an echo from 'Ode on a Grecian Urn'; if so, the phrase is not uncharacteristic of Green's habit of demanding poetry's support while effectively denying it the power to act. When Green directs his attention to the 'stinted expression of verbal signs' he stints the whole question too, by deploying gobbets of Shakespeare, Keats and Tennyson as simple emotional referents. It could be contended, in his defence, that, in stressing a dichotomy between 'heart' and 'sign', Green is most essentially Wordsworthian and that if we praise the scruple of Wordsworth's distinction between a 'mechanical' art and 'real and substantial' suffering we can scarcely deny our tribute to Green. It could also be observed that the question whether our 'feeling', i.e. 'faith', is more trustworthy than our reason, was raised by Sidgwick, who wrote of Tennyson's *In Memoriam* that he felt there 'the indestructible and inalienable minimum of faith which humanity cannot give up because it is necessary for life'. A pioneering study, in English, of Karl Barth's theology states that Sidgwick's gloss on Tennyson 'admirably catches Barth's tone' and that it also anticipates Barth's stress on 'the intolerable crisis which is felt when life is viewed "existentially"'. I would say that the comparison is too generous and that the citation of Barth makes one only too acutely aware of what is deficient in Sidgwick, and in Green also. The distinction between rigorous abnegation and easy abdication, so keenly asserted by them in principle, is not always marked in their own practice. When Wordsworth says, of the female vagrant, that:

> She ceased, and weeping turned away,
> As if because her tale was at an end
> She wept; – because she had no more to say
> Of that perpetual weight which on her spirit lay

116

he is indeed implying that words are 'in some degree mechanical' compared to the woman's action and suffering. But in order to bring out the difference Wordsworth puts in a collateral weight of technical concentration that releases the sense of separateness: the drag of the long phrasing across the formalities of the verse, as if the pain would drag itself free from the constraint. In 'as if' and 'because', pedantically isolating her, we glimpse the remoteness of words from suffering and yet are made to recognize that these words are totally committed to her existence. They are her existence. Language here is not 'the outward sign' of a moral action; it is the moral action.

In Green's work we may find an approximate understanding of the nature of such action, but we are indebted to his pupil, friend and biographer R. L. Nettleship for a precise formulation. Nettleship perceived that:

> the consciousness which we express when we have found the 'right word' is not the same as our consciousness before we found it; so that it is not strictly correct to call the word the expression of what we meant before we found it.

A. C. Bradley recalls that it was Green who suggested to Nettleship that he might 'approach philosophy from the side of language'. One cannot readily determine from the context whether 'from the side of' is a direct quotation from Green rather than Bradley's paraphrase. It is perhaps fair to say that either Green or Bradley, or both, regarded the matter as one of optional angles rather than as a necessary belonging to such a world. Green's comparative diffuseness on this issue is made all the more noticeable by his otherwise strong emphasis on necessary belonging, on 'the "I" that has felt as well as thought, and has thought in its feeling'. It is in a passage concerning the 'relation of each to the other' that Green comes closest, if not to a realization, at least to an adumbration, of Nettleship's perception:

> As has been pointed out, the sensible 'here' has, while I write it, become a 'there', the sensible 'now' a 'then'. We may call the sensible 'heres' and 'nows' an 'indistinguishable succession of points or moments', 'each chang-ing place with that which goes before'; but in the very act of naming, *i.e.* of knowing them, we transmute them.

If it were possible to isolate the final statement ('but in the very act of naming . . .') from the overriding context it would also be possible to argue that Green is moving towards Nettleship's position; but such a plea cannot be entered. For Green, the transmutation is into 'categories' and 'general attribute[s]' whereby we 'who are within the process' experience a form of 'placing ourselves outside the process by which our knowledge is developed'. Green stresses two major issues of this dualism: the knowledge so gained is 'imperfect, and, through its imperfection, progressive'; it 'implies the undivided presence of the thinking self'. It could be said that, while the pattern of Green's ideas about the thinking self is centripetal, the pattern of his thinking about ideas is centrifugal. There is therefore a dualism within the texture of what he makes, and this dualism divides what he does from what his admirers say that he does.

'[Green] gave us back the language of self-sacrifice', wrote Henry Scott Holland. That Green gave to his readers crucial ideas of self-sacrifice is beyond dispute:

> He may 'find no place for repentance' in the sense of cancelling or getting rid of the evil which his act has caused; but in another sense the recognition of himself as the author of the evil is, in promise and potency, itself repentance.

That these ideas were taken up and put into practice, to the benefit of the community, is again an incontestable fact. It seems to me, however, that Holland's words must be qualified. Green gave ideas but he did not give a 'language'; and in failing to give a 'language' he closed off one dimension of moral intelligence. The importance of Nettleship's perception is that it does not seal off the moral action from the linguistic action. Indeed it is the positive coalescence of these forms of action that is manifested in the cogency of his statement.

In the course of a valuable discussion of the traditions of moral thought, Mrs Dorothea Krook has drawn attention 'to the significant common ground that may be seen to exist between poetry and philosophy when both are viewed as products of the creative imagination'. It is in this domain that Green, who had so much to say, has so little to give. And this realization strikes with a particular irony. It was an Anglican priest, Dean Inge, who quite properly discerned that 'From Coleridge to von Hügel . . . the deepest and the most forceful [moral and religious] teaching has come from lay writers'. He included in his list the names of Green, Sidgwick and Nettleship. If 'deep seriousness and earnest desire to know the truth and make the world a better place' were the sum of the moral imagination, Green would be above criticism. But it is not; and, as one who, albeit unwittingly, divided morality from the moral intelligence, Green must, I think, be vulnerable to attack. His perorations are ominous, looking upwards and outwards to times 'when the poet shall idealise life without making abstraction of any of its elements'. Yet even while warning against abstraction Green failed in one necessary act of concentration. The 'element' was there and he could not see it. The extent of his loss, and ours, becomes clear when we turn, not only to Nettleship's statement, but also

to a discussion which his own argument seems, in some respects, to antici-
pate: Donald MacKinnon on 'introspection and the language of freedom':

> 'I could have done otherwise'. Such a phrase might express not simply a
> retrospective glance at what might have been, but *an act of repentance* which it
> makes concrete. When the prodigal said that he would 'arise and go to his
> father', he was no doubt, from one point of view, recording the fruit of inward
> communing with himself. But at the same time he was making that commun-
> ing something more than a mere daydream, and he made it so by linguistic
> action. He said something to himself, and, by saying, he did something.

We may perhaps say that, viewed in the light of this suggestion, Green's
work is both a summons to activity (including a summons to acts of sacrifice
and atonement) and also the outward sign of fruitful inward communing;
but not 'at the same time'. It is not a doing-by-saying. Consequently,
although he asserts the significance of the poetic act as a moral act, his
assertions are 'mere daydream'.

If Green, despite these several reservations, still attracts one's prolonged
and intensive meditation, it is chiefly because he bears witness, as in the
impressively characteristic 'An Answer to Mr. Hodgson', to that central and
inescapable conflict in which all who engage in public discourse become
involved. The original subscribers to Coleridge's *The Friend* numbered just
under four hundred. He was satisfied, we are told, 'to direct his remarks to
the "learned class" he was later to call the "clerisy"'. He required 'the
attention of my reader to become my fellow-labourer' but from the surviving
comments of several self-assured readers – Josiah Wedgwood, the Lloyds,
and others – it is evident that some of them considered that he asked too
much. His readers, with few exceptions, rebuffed his attempts, finding him
'abstruse and laboured' as others, later, found Green 'cruelly inarticulate'.

As Blake said, the accuser is god of this world. It is possible to find in this
recognition a necessary resilience: 'A man walks across this empty space
whilst someone else is watching him, and this is all that is needed for an act
of theatre to be engaged'. Such resilience is also present in the poetry and
prose of Wordsworth, especially in the 'Preface' to *Lyrical Ballads*, in the
poetry and marginalia of Blake and in the flashes of grotesque comedy, the
interpolated cries and groans, of *Biographia Literaria* and *The Friend*.
Green's lectures, as we know them in the expressive re-enactments of those
who attended them, make their own restricted but valid discovery of the
'formal engagement'. Green was, in a very immediate sense, requiring his
audience to become his fellow-labourer. Even if we say that he does little
more than demonstrate the pitch of attention at which the true (as opposed
to the spectral) Coleridgean 'clerisy' might be expected to work, that is

119

ample praise and makes a proper rebuke to the gods of this world. In the sacrificial nature of his perplexed persistence and in his vulnerability to the accusation which his servitude draws upon itself, Green achieves his own substantial freedom and power and remains as one of the most crucial writers in nineteenth-century Britain. There are triumphs that entrap and defeats that liberate. Green is creative in his distress. To speak of his exemplary failure is to see him in the light of a noble phrase borrowed from Forsyth; it is to say that he 'passed through negative stages to his positive rest'.

EIGHT

What Devil Has Got Into John Ransom?

The question, as it stands, is rhetorical and contentious. To Donald Davidson, who raised it, the answer was clear enough. It was a devil of a betrayal, as Ransom had accurately predicted. 'It will seem', he wrote, 'like treason and unfriendship' to Davidson's die-hard Southern Agrarianism not merely to recant from those once-shared principles and policies but also to flourish one's recantations in print. To Allen Tate, working a later tack and writing in a different vein, Ransom was not so much possessed of a devil as gripped by a 'mania' which drove him in later years to the ruinous rewriting, the 'compulsive revisions' of poems written in early middle age. Those who know their Ransom will justly observe that not all his revisions were in effect 'ruinous', though the impulse may have been; and his admirers will be inclined to add that 'devil' and 'mania' are weirdly ill-fitting masks for a poet whose favoured stance was a compounding of stoic ceremony and stoic laughter. It is true of course that laughter need not always be pleasant nor ceremony what it seems. Davidson once called Ransom's epistolary style 'meaty and flavorsome' and Tate with equal relish remarked that 'John throws a very wicked style, indeed'. Those were early days. Some years later Davidson would take exception to his 'cruelly polite snobbishness'. Would we be justified, then, in saying that Ransom possessed a devil of a style, in keeping with his own premiss that 'if we were better men we might do with less of art'? This would be the customary half-truth. Even so shrewd a stylist as Ransom does not entirely possess his style, self-possessed as he may seem; but neither is he wholly possessed by it. There was more than a mere tactical diffidence in his remark that 'it is common for critics to assume that a good poet is in complete control of his argument'; implying, of course, that he may well not be. In the writing of a poem, he says, 'argument fights to displace . . . meter' while 'meter fights to displace . . . argument'.

Here, though, my thesis may appear vulnerable, not merely to the stringent logic which Ransom preached and on frequent occasions practised, but to any barely audible murmur of dissent. When I speak of Ransom's style, I gather allusions randomly from poetry and prose. Judged by one of his own formal distinctions, this will not do. The author of 'a good poem', he once suggested, is in its making 'freed from his juridical or prose self'. In a late essay, however, he suggests an affinity between the 'vocalism' of the poem and that of 'meditative and imaginative prose'. I do not think that he makes his point at all well; but it is a concession and I intend to exploit it. One might add, with more fairness perhaps, that the 'juridical or prose self' can be taken as a figure of speech referring to the empirical, quotidian self which is to be distinguished from the 'I' of the 'poem' (whether in verse or prose). In saying this I knowingly allude to his celebrated definition of the poem as 'nothing short of a desperate ontological or metaphysical manoeuvre'. If it is permissible to define a poem in this way – and it does seem, I concede, a formula at once all-embracing and exclusive – then Ransom's essays can be defined as poems, since 'desperate ontological . . . manoeuvres' is what they are. Allen Tate declared that Ransom's critical essays 'have the formal felicities of his poetry' but this, though a crucial perception, remains a half-truth unless we add that they show the concomitant infelicities of clearing one's meanings at an extreme pitch of concentration and in circumstances of complex difficulty. If Ransom's essays are prose-poems they are so in a particular dimension; and it is a dimension not of freedom but of constraint. One could argue that there is a characteristic poetic crisis where rhetoric is ontologically determined (which is tolerable, even desirable) and where ontology is rhetorically fixed (which is intolerable but possibly unavoidable).

Ransom's definition, embracing 'agony' as well as desperation, occurs in the closing pages of a book which begins by rebuking the young poets for their neglect of the craftsmanlike qualities: 'wit and playfulness, dramatic sense, detachment'. Are we to see Ransom's art as a measure of these extremes? Some would say yes. A phrase – 'torture of equilibrium' – from one of his best-known poems is taken as an appropriate epigraph for his life's work. Miller Williams calls him 'the supreme equilibrist' and remarks that 'balance is the word which will probably surface most frequently in any discussion of Ransom's poetry'. Despite the consensus we may better appreciate Ransom's final achievement by not shirking the occasions when he is thrown off balance. In his fine dialogue-poem 'Eclogue' one of the characters declares 'We are one part love/And nine parts bitter thought'; and the nine:one ratio seems to me no less emblematic of Ransom's consciousness and craft than is the five:five equilibrium to which our attention is most often directed. Of course, one might be happier to express it as an equation and, indeed, Ransom, with his emphasis on such terms as 'playfulness', 'dramatic sense', 'detachment', seems to recommend that we should. The particular acidity of his 'playfulness' on the subject of Romantic suffering is characterized in the poem 'Survey of Literature' by his rhyming the name Shelley with the phrase 'pale lemon jelly', but he does not come off from the confrontation with much credit. Shelley is Shelley and Ransom is silly. What is it that persuades a poet of such intelligence to think so highly of so shallow a poem and to cling to it so obstinately while renouncing work of much greater quality? Why compound impertinence by injustice to oneself?

Impertinence is often a symptom of nervous strain; and 'nervous strain', according to Ransom, who credits I. A. Richards with the thought, is the 'peculiar plight of the moderns'. All commentators on Ransom's life and work have exhibited an understandable and proper reluctance to show the scars of the private man. Allen Tate wrote, after his death, that he carried 'an intolerable burden of conflict that only occasionally, and even then indirectly, came to the surface'; and Ransom himself, in old age, conceded 'an intense sympathy' for T. S. Eliot, 'that tortured soul who achieved serenity'. But what is this 'privacy' of ours, whether of the intellect or of the passions? George Eliot's dictum is well-known: 'there is no private life which has not been determined by a wider public life'; and though the main drift of Ransom's metaphysics could be called anti-deterministic he was enough of a Calvinist to find a number of painful conclusions inescapable. A poet of his calibre makes himself unique through 'an act of will' which is primarily an act of critical 'attention' and it would be doing him a disservice to imply a quality of uniqueness to any of his raw apprehensions as such: they are the Anglo-American cultural commonplaces of the last hundred years, from Henry James to the Lost Generation and beyond. 'The plight of a religious poet in an age of reason is desperate'; 'I am an Anglophile . . . But I am not so Anglophile as I am American'; these are the not-unfamiliar signals of the post-Romantic lost traveller. One of Ransom's poems, 'Old Mansion', is subtitled 'after Henry James' and another is called 'Man Without Sense of Direction'. It should be added that, in Ransom as in every writer of significance, context is everything and that he takes care not only to purge and enrich clichés by exactness and resonance but also to abrogate, by contextual means, that pathos which certain words and phrases might excite in isolation.

Donald Davidson once remarked that 'John isn't a balance wheel, he is a mystery', but 'mystery' itself is perhaps a term too much steeped in pathos. It is arguable that Davidson might more happily have kept the mechanistic bite of the metaphor if he had said: not a balance wheel but an eccentric. That which is eccentric is not concentric; it does not share a common centre; and phrases of absence, of exclusion, are a recurrent theme in Ransom's work, in the polemical and critical prose as well as in the poems. He broods deliberately, though, in terms of a lost focus: an idea which usefully involves the hazards of psychic distress with those of technical maladjustment. Of explicators and explications he remarks that 'respect ceases/For centers lost in so absurd circumference'. In 'Winter Remembered', the 'I' of the lyric narrative finds himself 'Far from my cause, my proper heat and center'. For the eponymous protagonist of another poem, '[Robert] Crocodile'. 'It is too too possible he has wandered far/From the simple center of his rugged nature'. An 'eccentric', it may be noted, is also a mechanical device which can convert rotary into rectilinear motion; and we could remark that, whereas the characters in Ransom's poems tend to go round in circles, he has acknowledged a liking for the 'workmanlike poetic line carrying forward the argument'. But, as we said, context is everything; and Ransom's 'liking', *in situ*, is a confession of some embarrassment, as he feels that a relish in the carrying forward of argument can be indulged at the expense of characters, who matter more. This is one check upon free-ranging hypothesis; and there are others, for example, his own sceptical playfulness with the term 'eccentric' itself. Commenting on Donne's 'Valediction Forbidding Mourning' he suggests that the 'metaphysical figure . . . tries to endear [the lovers] to us, by making them slightly . . . eccentric'. 'Old Man Playing With Children' is indeed a celebration of eccentricity and an indictment of 'middling ways'. There is a further complication in Robert Buffington's insistence that 'the three careers of Ransom – as poet, as critic, and as *Kulturkritiker* – run their courses about [a] common center'. If we follow Buffington's argument, the 'center' is itself a kind of denial: the determination to 'keep . . . reason and emotion separate'. This feat, I would have thought, is impossible – for anyone other than the victim of some form of psychosis, and certainly for the poet engaged in what Ransom calls 'the agony of composition'. One could conceive, however, that the desire could well be there, as the devil of a temptation.

Allen Tate, as scrupulous in his weighing of sympathies as of antipathies, traced back the 'mania' which, in his view, afflicted Ransom in his dotage, to an excessive 'reliance on *logic* as the ultimate standard of judgment'. 'Logic' is, like 'balance', a word which tends to recur in discussions of Ransom's work. In Rubin's recent study it is successively described as 'calm', 'inexorable' and 'remorseless'. It becomes increasingly difficult, though, to see how this 'logic' really works, for, as Rubin demonstrates effectively enough, when circumstances arose which made a 'system' imperative, practical 'logic' was in fact deficient. In the Agrarian symposium, *I'll Take My Stand*, published in 1930, not only the intrinsic coherence of the argument but also the effectiveness of its dissemination were vitiated by conceptual confusions and by errors of tactics and policy; and for these Ransom seems to have been in considerable measure responsible. To this evidence might be added Graham Hough's observation that Ransom's critical writings are by no means free from solecisms, 'eccentricities of nomenclature and description and history . . . that would certainly merit censure in . . . a Ph.D. thesis'. I do not think it unfair to suggest that, if by 'logical' we mean 'capable of reasoning correctly', then Allen Tate is more consistently cogent than Ransom; and if we mean 'sticking to the facts', then Donald Davidson has the more literal mind.

It is ironic to find words such as 'inexorable' and 'remorseless' featuring so emphatically in appraisals of Ransom's work because there is a sense in which he is deficient in straightforward conceptual rigour. He wrote, in 1924, that 'no art and no religion is possible until we make allowances, until we manage to keep quiet the *enfant terrible* of logic that plays havoc with the other faculties'. This 'logic', of which he is fairly sceptical, may with some appropriateness be termed a 'spectral' logic since Blake's Spectre is 'the rational power . . . [which] is anything but reasonable . . . of the divided man . . .' 'Anything in which reason and heart are separated has its Spectre'. This 'spectral' logic approximates to the logic of 'abstraction' which Ransom consistently opposed, arguing with a particular and detailed irony against that which he regarded as an arrogant 'Platonism' in aesthetics and science. In any account of his thinking it is necessary to distinguish abstract 'logic', which he did not admire, from 'logical structure', which he did – which indeed he saw as essential to the attainment of the 'beautiful poem' – as well as from that characteristic virtù which T. D. Young has called his 'natural inclination toward honest (and sometimes severe) judgments'. Yet while the intention of exegetes is, quite properly, to praise Ransom's honest severity and command of formal structure, they finally assert the perfection of a 'logical' strategy which is in fact far from flawless and which is, in any case, of doubtful efficacy in the making of poems or of that 'meditative . . . prose' which Ransom admits into the ambience of poetic effect. When a critic is able, without discernible irony, to praise the 'remorseless' logic with which Ransom attacks the 'ruthless' nature of modern science and industry it may be taken as a symptom of some confusion. If the one is 'predatory' why not the other? There may be some way of explaining this, but the matter is not even noticed as something to be explained. Ransom has said that 'we may hardly deny to a word its common usage, and poetry is an experience so various as to be entertained by everybody'. This remark cannot be wholly devoid of pessimistic irony. What at first glance seems a moral imperative ('we can hardly deny') may in fact be only a resigned acceptance of the coercive force of 'common usage' and the solipsistic free-for-all. These nuances, rich in themselves, do not sit easily with Ransom's unambiguous affirmation that 'the kind of poetry which interests us is . . . the act of an adult mind'. There is a disjunction too abrupt to be resolved by such terms as 'tortured equilibrium'. 'Desperate' is a word to which Ransom sometimes resorts; and he permits himself crucial statements which communicate a sense of vertigo rather than of balance. 'Images are clouds of glory for the man who has discovered that ideas are a sort of darkness' he writes, and there is nothing 'balanced' about this: it is a perception, and endorsement, of extremes. As he presents it, the ontological world (he took Kant as his mentor) is one in which we ought to prepare our minds to be 'perpetually disquieted'. He suggests that 'a man is impelled to take action when he finds himself uncomfortable, or in need, or lost, in some given "situation"'; and if he has any reservation about his mentor it is that Kant finally 'takes comfort' in the idea of the mind's existing 'wholly in the barren and simple world of the quantities'. To hope for this, Ransom argues, is 'to hope for a peace that is hardly to be distinguished from suicide, and not at all from mutilation'.

Ransom's word 'situation' is a revealing term. 'Piazza Piece', 'Lady Lost', 'Janet Waking', 'Two in August', 'Moments of Minnie', 'Parting at Dawn', 'Parting without a Sequel', *'Fait Accompli'*, 'Conrad Sits in Twilight', 'Prelude to an Evening', are poems of 'situation' where the nuances range from simple 'setting' to the fullest implications of being in a devil of a fix. 'The Equilibrists' ('Predicament indeed!') and 'Captain Carpenter' ('he fell in with ladies in a rout') narrate fatal games of consequences. The title 'Prometheus in Straits' ironically narrows the convention, limits expectation: Prometheus, not 'bound', but in 'awkward or straitened circumstances, a difficulty or fix'. The irony or ambiguity in these poems turns indifferently upon the outcome of decision or indecision. Whether the protagonists act or do not act, they appear equally maimed by bitter circumstance. 'Situation' is inescapable; 'stance' or 'attitude' is vital. Ransom clearly favours a ceremonious stoicism, the capacity, perhaps, to 'learn a style from a despair'. The phrase is William Empson's, and there is a degree of reciprocity, of qualified rapport, between these two poet-critics. *The Kenyon Critics*, a volume which Ransom edited in 1951, includes an essay by Empson, who had visited Kenyon College as a Fellow in 1948. T. D. Young's biography of Ransom reproduces a photograph of Empson taken in the company of Allen Tate, Cleanth Brooks, F. O. Matthiessen and Ransom himself. The words 'qualified rapport' must be emphasized. Ransom published a piece in the *Southern Review* (IV: 1938–9) with the title 'Mr. Empson's Muddles' and, as late as 1969, his approach, though broadly admiring, was still critical. As Christopher Norris has pointed out, the later study reveals that it was mainly the 'character of logical explicitness' which Ransom had come to mistrust in Empson's criticism, 'the reading of the poet's muddled mind by some later, freer and more self-conscious mind'. This is by no means the only occasion on which Ransom's 'wicked style' is directed against critical exegetes, whose activities provoke him, in 'Prometheus in Straits', to 'gestures not of deference'. The Prometheus in this poem is, by implication, the long-suffering fire-bringer, the divine spirit, the Paraclete, at best embarrassed and at worst betrayed by the activities of the exegete; a circumstance depicted in the subsequent poem 'Our Two Worthies' as a near-blasphemous travesty of the Christian apostolic succession. Primary sources are compromised by secondary material.

It is scarcely conceivable that Ransom could have been unaffected by his own 'straits' or have felt entirely unembarrassed by his own criticism. Hough suggests that a subdued but persistent hostility to academe prompted Ransom's bouts of 'obstinate intellectual waywardness'. One must distinguish between waywardness and error. I cannot think that any writer perpetrates an actual solecism in order to show his defiance, least of all Ransom, who had his proper share of *amour propre* and who valued the 'beautiful thing' too highly to get factual details 'oddly wrong' on purpose. I concede the probability of emotional discomfort, however, and the likelihood of its being exacerbated in a man to whom 'academic standing' was 'his bread and meat', who was always 'seriously concerned about his prospects in the academic world' and who was never less than insistent that 'he must have full credit for all of his efforts'. Ransom's claims for the 'anonymity' of achieved art make it seem tantamount to a breach of etiquette to refer to these things. Nevertheless I would accept Rubin's suggestion that, in the essay on 'Lycidas' with its title 'A Poem Nearly Anonymous', 'Ransom on Milton . . . is also Ransom *as* Milton, or Ransom on Ransom'. In such passages Ransom's criticism is neither 'pure' nor 'impure'. That is to say, it attends meticulously to its own procedures; its self-elected argument is not marred by palpable illogic; and yet, if set in a broader, more 'historical' perspective, its lines of approach appear both wilfully determined and 'oddly wrong'. His hypothesis that Milton wrote 'Lycidas' 'smooth' and rewrote it 'rough' thereby demonstrating his 'lordly contempt' for the traditional restraints of his art and exposing 'the ravage of his modernity' is a view that cannot be sustained against the evidence of historical scholarship. The 'lordly contempt' is not Milton's but Ransom's; and this arrogance of speculation is as substantial a part of his creative being as is the amiable unpretentiousness, the 'serenity' and 'gentleness' to which tribute is so often and so rightly paid. There is, of course, no blatant self-expression in Ransom's work. Allen Tate once said that 'self-expression' is a term that ought to be 'tarred and feathered', and I know of nothing in Ransom that would dissociate him from that view. But self-affirmation may be very different from self-expression, and can take anonymity as one of its forms. This distinction is drawn by implication in a passage in *The World's Body* where Ransom differentiates between 'two kinds of individualism', the one 'greedy and bogus . . . egoism', the other 'contemplative, genuine and philosophical'. But this is a contrariety which cannot be established by mere assertion. It must be maintained, heuristically, by an exemplary vigilance and distinction of tone which ideally cannot be faulted but which, in actuality, falters and errs at times, as it must. This in itself is enough to account for the 'nervous strain' to which both Richards and Ransom allude. The contemplative, which is akin to Schopenhauer's 'knowledge without desire', annuls or transcends, and thereby redeems, 'predatory and acquisitive interest' and possibly 'nervous strain' as well.

It is certainly open to suggestion that Ransom's particular technical achievement has been the conversion of 'strain' into 'pitch'. To be 'contemplative' is to achieve 'pitch of attention'. Like the lover in an early poem by Robert Graves, whose work he admired and who was, in turn, one of his first champions in England, Ransom hears 'the groan of ants who undertake/Gigantic loads for honour's sake'; but he can also hear the 'minute whispering, mumbling, sighs' of language itself. As Vivienne Koch has sensitively perceived, Ransom's alteration of the line 'two immaculate angels fallen on earth' to 'angels lost in each other and fallen to earth' not only 'enriches the denotative content'; it also strengthens the poem's logical structure. In their first appearance in the magazine *The Fugitive* (March 1925) two lines of 'Eclogue' are both mannered and fey ('They run to the fields, and apprehend the thrushes,/And print the fairy dew . . .'). 'Apprehend' is conceitedly stiff and 'fairy dew' is tawdry. In *Selected Poems* (1945) this becomes 'They run to the fields, and beautiful the thrushes,/Fabulous the dew'. In changing 'apprehend' to 'beautiful' Ransom exchanges a self-assertive intellectualism for an undemanding banality. But, in the end, the banality is not there; it has been dissolved in the uncanny rightness of the word 'fabulous'. This is a dangerously volatile word but Ransom has tamed and tuned this very susceptibility, allowing the halftones of 'not really there' and 'miraculously *there*' to blend most 'beautifully'. Ten years later, in *Poems and Essays* (1955), the lines read 'And listen! are those not the doves, the thrushes?/Look there! the golden dew'. The 'golden dew' of 1955 is no less a cliché than was the 'fairy dew' of 1925. It is not revision but reversion; not renunciation but ruin. What has 'impelled' Ransom; or what has led him on? In the three versions under consideration, Jane Sneed's vision of the fields of dew is immediately challenged by her lover John Black, who has the 'last say': 'O innocent dove,/This is a dream.' Does Ransom believe that he has made the 'dream' so 'fabulous', so miraculously 'there', that we might be too dazed to attend to John Black's reproof unless he turns syntax itself into a kind of travesty? Or is he going against his own best advice in ignoring the 'disparity' between a poem's 'grammar' and its 'real logic'? The 'grammar' of the dialogue between Jane Sneed and John Black is more deliberately spelt out in 1955 but the 'real logic' may belong more to the 1945 reading where miraculous simple truth is a 'dream' far worthier of John Black's renunciatory cry than are the fussy expostulations of 1955. The battle is not yet over, however. The third edition of *Selected Poems* (1969) contains a further variant; and the analysis which I have so far offered is substantially qualified by this. The whole of the stanza in question is now omitted, the antepenultimate stanza moves into the penultimate position, and John Black's 'This is a dream' now refers to the previously unconsidered claim that lovers' eyes 'are almost torches good as day,/And one flame to the other flame cries Courage,/When heart to heart slide they.' And in the light of this we look again at the lovely stanza of 1945 and find that it is a succession of lyrical arabesques around the keynote 'unafraid' (which is itself a close variant of 'courage' two lines before. Ransom may have felt it was too close). F. O. Matthiessen suggested that Ransom, in the act of revising, 'felt his way back into the meaning' of a poem. This may now seem a truism perhaps; but it is originally perceived and sensitively phrased and stands up well to an empirical test. There will be legitimate differences of opinion about these things; but I am persuaded that Ransom, through a process of apparently marginal tinkering and trimming, discovered a long-concealed central disproportion in the poem and finally, after something like forty years, found it possible to clear his meaning. If his own phrase, 'torture of equilibrium', can be said to apply to his own work (an application which I would strongly qualify but not wholly discount) then its validity is surely tried here, in the 1969 text. The hurt and the poise are exacerbated and enhanced where the 'dream' to be revoked is of 'courage' and tenderness rather than of 'thrushes' and 'fabulous . . . dew'. The 'real logic' of the poem belongs, after all, not to the 1945 version but to the text of 1969.

Ransom once remarked that poets have to 'consider technique as a charge upon their conscience' and this is burden enough; it is a devil of a conscience. To write at Ransom's pitch of attention is to be in ontological and semantic 'straits'. Acute sensitivity to the richness of 'denotative content' is also vulnerability to acoustical din, a situation in which the very finesse of the poet's perceptions becomes a source of bafflement and panic and, at worst, of self-destruction. In his chapter on I. A. Richards in *The New Criticism* Ransom wrote that 'the confusion of our language is a testimony to the confusion of the world. The density or connotativeness of poetic language reflects the world's density'. This statement could fairly be said to come from the heart of Ransom's metaphysics and to go to the heart of his poetics. It can scarcely be denied, however, that if 'world's density' is a metaphor, 'confusion of . . . language' is a fact. Though syntax may shape an antithesis, imbalance will not melt into equilibrium. In attending to this, however, we give attention to the 'crucial' Ransom: to the method and the 'madness'. My justification for this last piece of impertinence is a recollection of his own observation that 'Platonic . . . idealists . . . who worship universals, laws . . . reason . . . are simply monsters'. His own idiom is not

129

devoid of Platonism, as his recourse to such terms as 'good poet', 'good poem' and 'beautiful poem' may sufficiently indicate. To insist on a 'grammar' of total prohibition when the 'real logic' of the situation points to relative acceptances is itself a 'monstrous' idealism or, as Allen Tate said, a 'mania'. This is not simply a word which is being turned upon Ransom. Whether or not Tate recalled the fact, the word already belonged to Ransom's weaponry of wisdom and satire. He once described 'an absorbing speciality' as 'a small mania' and quoted, in a different context, T. S. Eliot's definition of a sub-species of Jonsonian 'humour': 'a simplified and some-what distorted individual with a typical mania'. If we ask whether Ransom's infelicities signify a 'humour' or a 'negative capability' it is a question which we are bound to try to resolve in due course. It may be observed, in the in-terim, that the irony of Tate's trapping Ransom inside his own armour is not particularly unhappy. It is a timely reminder of the closeness, in agreement and dispute, between friends and colleagues who formed the 'Fugitive' group, later the nucleus of the 'Agrarian' movement. Donald Davidson's terse evocation of 'great mutual attraction, with perhaps some repulsion here and there' will serve admirably. In the 'Acknowledgements' to *The World's Body* (1938) Ransom writes of his 'obligations' to Tate and of observations surfacing 'in a manner to illustrate the theory of anonymous or communal authorship'. And it is from a poem published in 1969 by another former student and life-long friend of Ransom, Robert Penn Warren, that I take the injunction: 'In this century, and moment, of mania,/Tell me a story'. If, going against the grain of Ransom's principles, we were to generalize upon his work and make abstraction from his dense particularity, we might say that his poems are indeed 'stories' and that some of his best deal with 'moments of mania', extraordinary delusions about the nature of love or wisdom or the capacity for endurance. If constrained to synthesize the diversity of his aesthetic and polemical pieces, we could argue that, from the first volume *God Without Thunder* (1930) to the last collection of occasional pieces, *Beating the Bushes* (1970), he testifies against a 'century of mania', the epidemic excitement of the 'naturalistic rabble' and the 'aggressions', the 'cold fury' of science. But however strongly and however rightly we maintain that Ransom's poems are not 'confessions' but 'stories' and that his essays are forensic speeches of prosecution and defence in which admission is, by convention, oblique, we must still concede that the double consequence of a poet's involvement with language is complicity and revelation. Exegesis, which is a kind of science, is in a continual 'cold fury' of appetite for proof of absolute 'mastery', but the poet's 'agony' is more likely to be an empirical involvement with that process whereby 'confusion' becomes 'density'. But is it, perhaps, only 'empirical' in theory? If Ransom's 'equation' is in fact unreal, 'transformation' is a leap of metaphor not a process of reasoning.

Ransom, I think, concedes as much. 'Poetic metaphor' is a 'loose analogical technique', he says; and, once we have conceded analogy, we can 'approve' the conclusion that the 'ritual of the public occasions' is a style of poetry and encourage the idea of 'ceremony' as an interchangeable term. Yet here again logic is baffled. Public occasions may be poetry but the poem, or, more precisely, the modern poem, is not a public occasion. 'It is not about "res publica," the public thing'. We seem to detect here a tincture of irony, the mild private travesty of a gross public indifference. Allen Tate puts it more candidly when he numbers poetry among those 'discredited forms' which have been 'defeated by the popular vote'. Ransom's observation is, in contrast, a 'kind of obliquity' which, in his view, art so often is. One might say that art is 'oblique' because other things are 'unequal': 'We are creatures of a knowledge which is hopelessly unequal in its features' he wrote in 1930, a statement which the young Allen Tate had anticipated by five years in one of his pioneering informal appraisals of Eliot. Modern poetry, Tate said, is not 'intrinsically difficult' but poet and audience are trying to effect an exchange in a 'currency of two different languages'. Ransom greatly admired 'strong-minded' Tate. He admired creative and critical tough-mindedness generally, though preferably disciplined by the traditional courtesies, and was himself inclined to stoicism. If he was 'shaken' it was assuredly 'not as a leaf'.

Since 'irony', together with 'logic' and 'equilibrium', is a term consistently favoured by Ransom's critics we should heed his own caveats. Irony is acceptable in the particular circumstance; it is 'objectionable only when it is made poetic staple'. If we found that works of art 'were agreed in never exhibiting moral aspirations . . . or human efforts triumphant . . . it would occur to us that artists were professionally mean and misanthropic, and fixed their evidences'. In his own work Ransom displays a decent modicum of 'human efforts triumphant', not only as matter for narrative (in the poems) or demonstration (in the critical prose) but also as a staple quality of the rhetorical tone. Poetry's relation to 'res publica' can never be other than tangential; but this can be seen as but one aspect of a more profound and far-reaching 'contingency' which it is man's glory, as well as his fate, to encounter. There can be no question but that the term 'contingent' and its cognates have formed a constant ingredient in his verbal stock for more than forty years, making his style of metaphysics 'meaty and flavorsome'. In *God Without Thunder*, 'nature still waits, unconquerable, unintelligible and contingent'; in his second prose book, *The World's Body* (1938), he points to 'things as they are in their rich and contingent materiality', 'the world populated by the stubborn and contingent objects'. In an essay of 1945, later collected in *Beating the Bushes* (1970), we read of 'the vast depths of . . . natural contingency'. Closely correlative idioms include 'the stubborn

131

"manifold" of this world', 'marvelling, and revelling in the thick *dinglich* substance', 'the denser and more refractory original world', 'ontological density', 'the dense and brilliant yet obscure world of the modern poets', 'stubborn substance', 'dense natural context'. This terminological congress, this feast of concepts, this 'thick *dinglich*' spread, should be sufficient to dull the edge of anyone's anxiety about the perils of 'abstraction'.

F. P. Jarvis has argued for the influence of F. H. Bradley on Ransom's critical theory, but, leaving influence aside, I am more struck by affinities with the other Bradley, the author of *Shakespearean Tragedy*, whose thinking, like Ransom's, was considerably directed by his reading of Kant and Hegel. I am even inclined to suggest that Ransom unlocked his metaphysics of poetry with the same key that Bradley used to unlock his metaphysics of tragedy: both moving Hegel's term 'ethical substance' backwards and forwards until it 'clicked' with their own thought. Bradley embodied it in his definition of 'the essentially tragic fact'; Ransom quotes it in a late essay. In 'Hegel's Theory of Tragedy', a chapter in his *Oxford Lectures on Poetry* (1909), Bradley writes of the 'self-division and intestinal warfare of the ethical substance, not so much the war of good with evil as the war of good with good'. Ransom's first publication, at other than a parochial level, an article called 'The Question of Justice', appeared in the July 1915 issue of the *Yale Review*. Ransom argues here for the 'right' and 'moral dignity' of both Britain and Germany and claims that the tragedy of the situation is that 'two good ideals should prove so irreconcilable'. It almost seems as if the accident of 'intestinal warfare' in the 'denser and more refractory world' of 1914 enabled Ransom to make a literal translation of Bradley's Hegelian metaphor of 1909. But if we are sufficiently tempted by the approximation to allow ourselves the supposition that Ransom was acquainted with the Oxford Lectures, the logic of the hypothesis requires his knowledge of Bradley's other statement: '[Poetry's] nature is to be not a part, nor yet a copy, of the real world (as we commonly understand that phrase), but to be a world by itself, independent, complete, autonomous'. There is a footnote to *The World's Body* which states that 'the artist resorts to the imitation because it is inviolable, and it is inviolable because it is not real'. Bradley's 'as we commonly understand that phrase' is a timely reminder that our perplexity is caused less by ontological definitions *per se* than by the 'currency of two different languages' which, regrettably, happen to share the same vernacular. But Ransom also says, in the same note, 'As science more and more completely reduces the world to its types and forms, art, replying, must invest it again with body'. How an art that is 'not real' can invest the real world with 'body' is a matter which Ransom never resolves, in its juridical or prose sense, though he is at pains to do so by 'sly analogy' (the epithet is his) 'taking advantage of the hazard of language'. As a type of pseudo-logic, 'sly analogy' is more dubious than the typical metaphysical conceit, which plays the game properly: but we must accept Ransom's attachment to it because, five years later, he is still in its grip. 'A poem', he notes in 1943, 'densifying itself with content ... is about the dense, actual world'. We in turn are bound to observe that Ransom's verbal substance is at its thickest and most suggestive where its 'logic' is thinnest. 'Is about' is a pathetically frail locution to act as the ontological membrane between two such *'dinglich'* clauses. If we are to have pseudo-logic, Marvell's 'A Definition of Love' is preferable. The argument presented there, that the more 'truly parallel' things are, the less likely it is that they will ever meet, is in effect saying 'don't design "perfect" analogies as a substitute for "identity" because, in the end, they won't "come across"'.

Ransom once pertinently remarked that 'the seventeenth century had the courage of its metaphors' and this stands as one of his most telling appraisals. At his best he is himself a metaphysical poet in this sense, deriving the proper power of his tropes from one of the most traditional of all metaphors, the Hebraic *ruah*, the rushing of a mighty wind, the Holy Spirit, the creative voice which, in one of his last essays, he affirms is also the voice of the poet. In 1930, in his first prose book, Ransom said that 'ruah' appropriately represented 'majesty combined with contingency' and I believe that he demonstrates the courage of his own metaphors when he allows the old conflict of form and substance to be just that. 'Trope' is a term possibly over-resorted to in recent years. I use it because it means 'turn', and it is a particular kind of turn that I have in mind. Ransom's own definition of 'trope' is apposite. 'Figures of speech twist accidence away from the straight course' is a precise description of the little swirl, or gyre, of stubborn reiterative outcry, disturbing but not halting the ceremonial procedure of his traditional verse-forms: 'Why should two lovers be frozen apart in fear?/ And yet they were, they were'; 'And heard the clock that clanged: Remember, Remember'; 'No, No, she answered in the extreme of fear'; 'A cry of Absence, Absence, in the heart'; 'Honor, Honor, they came crying./ Importunate her doves'; 'Whose ministers post and cry, "Thus saith, Thus saith"'; 'until I knew/That I was loved again, again'; 'Conrad, Conrad, aren't you old/To sit so late in a mouldy garden?/ ... Anchovy toast, Conrad!'. These little irruptions of outcry may be encouraging or censorious, summoning to tea or laying down the law in some other way; or they may be passionate protests at enforced obedience or desperate affirmations of disobedience; but they are so human in their ring (even if given to 'doves' or a 'clock') that the sound of the kites, in Ransom's most celebrated poem, repeatedly whetting their beaks 'clack clack' over the poor remnants of Captain Carpenter, is like a disgusting parody of the human activity. In itself, of course, there is nothing unique in such repetitive exclamations. In lyric poetry instances from Sidney, Keats and Hopkins spring immediately to mind; but the particular interest of Ransom's reiterations is threefold. Firstly they are, as he says, 'importunate'; they irrupt into preordained ceremonies (i.e. the 'forms') or into reverie or dull habitude (i.e. the narrative occasions) which would be easier to conduct, or to bear, if they did not. Or they speak remorselessly, in moments of crisis, of the normative, the 'code', that has been outraged or forfeited. Secondly, they occur often enough and in such a manner for us to call the pattern 'characteristic'. Thirdly, this 'characteristic' trope in turn characterizes Ransom's feeling for 'the stubborn "manifold" of this world'. In an essay of 1942 he projects a notion of laughter which, 'if it could be articulate, would be found reiterating, Substance, Substance ...' at the 'ingenuous idealist' and his 'program'. So laughter shouts 'substance substance' at the abstract with exactly the same cadence as the heart cries 'Absence, Absence' or the importunate doves their 'Honor, Honor'. And cadence itself, at this pitch of intensity, becomes a form of substance, a monad. Giordano Bruno, from whom Leibnitz derived the term, conceived of the 'monad' as being at once a 'material atom' and an 'ultimate element of psychic existence'; and it is such a twofold sense which pertains here. It has a core of pragmatic truth applicable to the work of good poets in general and to Ransom's work in particular. The word 'cry' and its derivatives form what is perhaps the most frequently recurring group of words in his poetry; and the 'cry' is peculiarly a 'monad' in which the physiological 'speaks out' for the ontological. It is the minimal utterance which, paradoxically, is expressive of a multiplicity of psychic nuance: abandonment, affirmation, solitude, communion; as the medieval mystics perceived in their teaching on the efficacy of monosyllabic prayer. The particular characteristic of the reiterated cry is arguably its power to transform pure spontaneous reflex into an act of will. It is not without significance that the passage from *Hamlet* which Ransom chooses for analysis is Claudius' speech in which he breaks into his own instructions for the 'close following' of Ophelia with the reiterated 'O Gertrude, Gertrude!' It may fairly be inferred that Ransom's interest is caught by the sudden compounding of guilt, anxiety, grief and love in the most elemental utterance as it seeks to make desperate affirmation out of mere recoil. Ransom wrote with discriminating admiration, in *God Without Thunder*, of Psalm 88, praising its 'realism', its 'sense of hard – very hard – fact' that is yet 'not quite submissive to the fact'. If we were to seek a method whereby an ontological principle is to be transformed into an idiomatic monad, this requirement might be met by the idiosyncratic 'cry' of a Ransom poem, since a cry is simultaneously a 'fact' and the resistance to a fact, a 'reaction' and a reaching-out. Despite his avowed allegiance to 'classical' criteria, Ransom is not surprisingly re-enacting a typical Romantic 'situation': one that had been confronted by Keats in his Letters where, in February 1818, he appeared satisfied to cite 'a few Axioms' concerning poetry and, three months later, felt bound to declare that 'axioms in philosophy are not axioms until they are proved upon our pulses'.

When we read C. H. Sisson's finely judged appreciation of Thomas Hardy's so-called 'awkwardnesses' – 'The rhythm of the verse, with its hesitations, sudden speeds, and pauses which are almost silences, is the very rhythm of thought' – we understand how the implications of the Keatsian 'pulse' can be comprehended as 'rhythm'. When we read Ransom, between whose work and Hardy's several critics have suggested close affinities, the technical issues are a little different. Technically he is one of the least 'awkward' of poets and yet his formal grace is in a constant state of alertness against 'awkwardnesses' which, even so, contrive to irrupt into the manners and measures of the verse. I shall call this the 'trope of infelicity'; it is not infelicitous. It is, moreover, Ransom's characteristic development of the Keatsian 'negative capability'; characteristic in that its frequent burden is the articulation of some 'mania' in men and women, some intense solipsistic passion, in themselves the very opposite of the Keatsian virtù. Such an appraisal of the poems, however, leaves unchallenged Tate's description of the old Ransom's persistent interference with his own poems as 'mania'. What I have tried to show is that the 'mania' is the dark side of that coexistence of 'majesty' and 'contingency' which, in its other aspect, produces the masterly fables of the poems. It is a radical miscalculation, which Ransom obliquely concedes in his note to his most ruinously reconstructed poem, 'Prelude to an Evening', when he wryly remarks that 'the social issue is saved, but I, like some of my friends, am not sure whether an expiation is always in the interest of a fiction'. And I would say that Robert Buffington is wholly correct in his suggestion that the error stems from Ransom's confusion of theology and morality. 'Theology' in this instance equates with 'fiction' of course, and the 'devil' that has got into John Ransom at this late stage is a decent little devil whose highly moral whisper is that one can genuinely make amends for some actual or supposed betrayal in 'real life' by tampering years later with the evidence of the fiction. Whatever Ransom may have thought or said, the revised 'Prelude to an Evening' is a 'weakened and trivial form'.

That last phrase is from Raymond Williams's acute analysis of the failures of later Eliot. And when Williams goes on to suggest that 'it is part of the real complexity of this extraordinary man [Eliot] that he could repeatedly and genuinely mistake ... compromise for communication', I am inclined to note how effectively these words describe Ransom's own mistakes. The causes are not precisely the same, perhaps. Eliot could genuinely mistake 'the fashionable requirements' of Shaftesbury Avenue or the Edinburgh

Festival for the needs of 'a wider audience'. Ransom did not have to endure that degree of worldly temptation, but I do not think that there was any great difference in kind, despite the gross inequality of their respective worldly rewards. In each case the error is traceable to a root cause in the 'currency of two different languages' circulating within the same vernacular. It is at the same time a simple confusion and a deep intractable problem. One could say that Ransom diagnosed it perfectly well when he wrote that 'behind poetry is the whole loosely articulated body of the language'. The 'literal sense' of 'language' is 'the whole body of words and of methods of combination of words used by a nation, people, or race'. It is the 'loose articulation' of its figurative senses which, I believe, increasingly betrays Ransom. It may be agreed that in such locutions as, say, 'the language of polite letters' more is implied than 'the methods of combination of words'. Such a 'language' is in fact a 'code' of acceptances, a consensus of taste or prejudice. I would suggest that, in his later writings, Ransom is inclined to confuse 'consensus' with 'language' and the consensus is increasingly that of the 'dull readers – with whom I concede my own affiliations'. If this is not serious it is cynical, and, at the risk of sounding like Matthew Arnold's little niece, I cannot see that it is 'wholly serious'. More and more frequently, from *The New Criticism* (1941) onwards, Ransom is provoked into the writing of scenarios in which 'those Yale professors' absurdly confront 'the common barbarian reader'. The provocation is understandable enough. He testified, in *The World's Body*, to 'that terrible problem . . . of poetic strategy' and again, in *The New Criticism*, to his belief that 'poetic strategy' is 'the last and rarest gift given to poets'; and it is arguably a chastening experience for any innovatory writer to find his arduous strategy become the fluent tactics of his disciples. Ransom had attempted in 1938 to demonstrate a crucial 'paradox': how 'the possibility of poetry depended on an event that carried also the possibility of its destruction'. It is not unreasonable that in 1952, reviewing a new book by Cleanth Brooks, he should remark that paradoxes 'are easy to find once you are searching for them'. He is endeavouring, we may say, to disembarrass the 'real logic' of poetic strategy from the 'grammar' of methodology. We may say also that this is 'the essentially tragic fact': the 'self-divisions and intestinal warfare' within a body of critical writing 'more intense than a language has ever known'.

I have suggested that the double-consequence of a poet's involvement with language is complicity and revelation. I have also suggested that Ransom's essays are, like his poems, 'desperate ontological . . . manoeuvres'. There is a thrill of 'mastery' in the tropes; but it is possible for a poet to serve the integrity of thought and language in the exemplary nature of his constraint. The particular interest of Ransom is in the achievement (the revelation) and in the price he paid for it (the complicity). 'The bad artists', he writes, 'are cruelly judged . . . and the good artists may be humorously regarded, as persons strangely possessed'. This is Ransom being slightly 'mischievous'. His colleagues seem to have liked this streak in him; but I would suggest that his somewhat coy indulgence of the 'humble reader' does the integrity of his criticism a mischief. In *A Survey of Modernist Poetry* (1927), Laura Riding and Robert Graves had championed Ransom's 'Captain Carpenter' against the demands and prejudices of the 'plain reader' who 'does not want a critical attitude' but rather 'poetical feelings of simple pleasure or pain'. Nothing has happened in the intervening years to make one wish these words unsaid; and Ransom's sorry wit can be distinctly inferior to his grasp of infelicity. He has been led, like Eliot, genuinely to mistake compromise for communication. It is not only the 'bad artists' who are cruelly judged. The good are too. I do not think that Ransom would have wished it otherwise.

NINE

Our Word Is Our Bond

And I might mention that, quite differently again, we could be issuing any of these utterances, as we can issue an utterance of any kind whatsoever, in the course, for example, of acting a play or making a joke or writing a poem – in which case of course it would not be seriously meant and we shall not be able to say that we seriously performed the act concerned.

Sir Philip Sidney, as shrewd as he was magnanimous, evidently had it in mind to keep poetry out of the courts: 'Now for the *Poet*, he nothing affirmeth, and therefore never lieth.' Since his time poets, with more vision than apprehension, have been endeavouring to thrust it back into the theatre of litigation. 'The Imagination may be compared to Adam's dream – he awoke and found it truth.' 'The poem is the cry of its occasion, / Part of the res itself and not about it.' Ezra Pound, who wrote magisterially on behalf of poets 'All values ultimately come from our judicial sentences', in the end found himself cruelly and impotently at odds with the u.s. judiciary. Seeking to attribute causes, or even reasons, for this savage contretemps one is presented with several possibilities. It may be that Pound misjudged a critical matter of status or perhaps he misconstrued a fine point of semantics. A further complicating factor is that legal distinctions may themselves be classifiable as 'fictions' or 'peculiar paradoxes'. Pound's sense of 'justice' and 'value', we may feel, accords with an antique hierarchy of ideal ontological principles and agencies formerly defined in such terms as *intellectus agens* or *recta ratio factibilium*, whereas his indictment for alleged treason took place in the domain of *mens rea* and *actus reus*, the world of common law. He is vulnerable to accusations that he naïvely or wilfully regarded his wartime broadcasts as being in some way traditionally privileged and protected by his status as poet, 'boasting of the sanctity of what [he] carried'; an attitude at best archaic and at worst arrogantly idiosyncratic; oblivious of, or indifferent to, the 'real world' which lies 'out there', where things (and people) regularly 'get done'. This 'real world' of current documentary appeal reminds one in its simplicity of the 'substratum' of the early empiricists. Other forms of empiricism, however, are rather more formidable. 'If you are a judge and say "I hold that . . ." then to say you hold is to hold; with less official persons it is not so clearly so; it may be merely descriptive of a state of mind.' Tested against this distinction Pound's 'judicial sentences' begin to appear less magisterial. J. L. Austin remarks that 'a performative utterance will . . . be *in a peculiar way* hollow or void if said by an actor on the stage, or if introduced in a poem, or spoken in soliloquy'. In his sense of the term, 'ordinary language' is 'rich and subtle' in certain kinds of 'pressingly practical matter' but these matters exclude such 'parasitic uses of language' as 'Goe, and catche a falling starre' which are 'not serious', not the 'full normal use' of the medium.

Isaiah Berlin notes that Austin was 'determined to try to reduce whatever could be so reduced to plain prose'. Austin himself stated that 'accuracy and morality alike are on the side of the plain saying that *our word is our bond*', nonetheless making play with the motto of the Stock Exchange as he did so. Observation will incline us to feel that the timbre of 'plain prose' is unlike that of 'plain saying'. The former is simply reductive ('reduce . . . reduced'), the latter amply assertive ('alike . . . on the side of'). If we stir the soil about the roots of either of these locutions we unearth seventeenth-century shards. The plain intention of both statements to be directly functional does not obliterate such discoveries. The crucial questions are how much 'play' remains in language after the logical excisions have been performed and whether this play is definable as 'controlled interplay' or as that 'play' which means that something designed for precise mechanical utility is showing signs of malfunction. Austin's belief that the expression 'I do' occurs in the Anglican marriage ceremony is manifestly of some significance in a work entitled *How To Do Things With Words* even though the mistake may be, as J. O. Urmson avers, 'philosophically unimportant'. It at least provides a point of contact between two observations which would otherwise be hard to reconcile. The first is that in Austin's teaching 'there appeared to be nothing between him and the subject of his criticism or exposition'; the second attributes to him an awareness of 'the complex and recalcitrant nature of things'. The first is a bright memorable impression of a transparent success; yet, as the second implies, the very idea of a 'transparent' verbal medium is itself an inherited and inherent opacity. Where there is 'semantic content' it is most likely that there will be semantic 'refraction', 'infection' of various kinds; for example 'even "ordinary" language will often have become infected with the jargon of extinct theories' or 'considerations . . . may infect statements'. Once you have released such a word as 'infect' how is it to be contained? Systems that appear 'perfectly neat and easy', as T. H. Green said of Hume, have a way of raising 'awkward questions'. Empiricism admitted complexity. Locke, for instance, 'distinguished clearly between simple ideas and complex ideas'; but to allude to the 'complex and recalcitrant nature of things' is to hit upon a timbre quite different from the tone in which one refers to the aggregates and compounds of Empiricism which, though it did not entirely reserve such terms as 'manifold' and 'inextricable' for the errors and disputes of the old Schoolmen, nonetheless regarded the 'inflexible natures . . . of things' as amenable to rational understanding and analysis. Locke's opinion was that 'things themselves', if approached without 'preconceived notions', 'will shew us in what way they are to be understood'. That there is a marked difference in weight between 'natures' and 'nature' in these citations is itself a ponderable suggestion. The 'natures of things' are their compounded properties, the 'nature of things' seems to take on a metaphysical denseness.

Empiricism, we may remind ourselves, regards words as being 'of excellent use' when rightly understood and as powers only in a dark and negative sense when they 'impose on the understanding'. Berkeley's eschewing of 'all controversies purely verbal' is, in part, a caveat against mistaking 'empty words' for 'legitimate concepts', but it is also, in part, a verbal gambit, a bid to score in a controversy. The empiricists are in at least one respect heirs of the Cambridge Platonists. 'If it were not for Sin, *we* should converse together as *Angels* do' said Whichcote. The empiricists, too, would in theory be happier with forms of sensuous or conceptual communion, 'naked, undisguised ideas' 'stripped of words'. In practice, of course, their own words have pith and spice. Santayana's reference to Locke's 'controversial relish' is apposite, and Berkeley has something of this 'relish' too. Locke, in his *Conduct of the Understanding* (published posthumously, 1706), refers to a 'jejune and dry way of writing . . . a sour and blunt stiffness tolerable in mathematicians only'. In Berkeley's *Commonplace Book*, kept between 1705 and 1708, we read: 'the short jejune way in mathematiqucs will not do in metaphysiques & ethiques'. Here Berkeley's use of 'jejune', despite a first impression of affinity and concord, makes, I would conjecture, a 'retort' upon Locke, and essays raillery. The bite of the raillery is more strongly appreciated if we know that Locke's 'sour and blunt stiffness' is itself a wry tribute to that manner of writing, not a denigration of it. He distinctly prefers 'plain unsophisticated arguments' to 'plausible discourses'. Yet this paragraph of oblique regard is itself in no sense plain and unsophisticated. It manifests a plausible urbane wit that is pleased to appear dry and dour. In Locke's own style, and in Berkeley's rallying to his style, there is a briskly nuanced play which, slight as it is, is nonetheless in excess of the empiricist desideratum that words should excite only the 'proper sentiments', that, in the 'ordinary affairs of life', they should be received as no more and no less than acceptable tokens of exchange, 'signs or counters', 'fiduciary symbols'.

Seventeenth-century philosophical antagonists, such as Hobbes and the Cambridge Platonists, were united in arguing that plain speech had a status superior to that of metaphor or carefully-elaborated discourse, though it is necessary to add that Whichcote's 'candor' does not plainly equal Hobbes's 'Perspicuous Words' and that, at the heart of the matter where Hobbes and the Platonists pursue their most immediate concerns, plain speech has scarcely more status than has figurative speech. If it is preferred it is because it is supposed to get in the way less. The focus of Hobbes's attention is not words but power. The Platonists' concern is less with 'fiduciary symbols' than with faith, conceived of as human reason directly intuiting the Divine Reason. There is not, in either case, a neglect of the serviceable parts of language. Whichcote is always concerned to 'speak accurately' and Hobbes's audacity is as much a matter of style as of thought; but these immediate concerns are subject to overriding considerations and objectives; in the one case a keenness of withdrawal from the 'troublesome multiplicity' of the things of this world and in the other a demonstration of the world's ineluctable necessity. This, then, is the direction in which impetus, or entropy, impels us. Wordsworth's 'real language of men' is, like his 'native and naked dignity of man', still a fiction, still one of those 'airy useless notions' that Locke contrasts with 'real and substantial knowledge'. To employ another Lockean idiom, real and substantial knowledge 'bottoms' on real estate and common law, the basis of what T. H. Green sums up as 'the wisdom of the world'. Or, to choose a form of words less tinctured or 'infected' by one's own preconceived notions, we could perhaps say that neither Hobbes nor Cudworth would dispute unduly Austin's statement that we use 'a sharpened awareness of words to sharpen our perception of, though not as the final arbiter of, the phenomena'.

If 'real' is opposed to 'ideal', Austin, who admired the methodology of the natural sciences, the 'patient accumulation of data about actual usage', is a realist, drawing numerous examples from law reports, relating others to 'the conduct of meetings and business'. His work on 'descriptive fallacies' has been of particular utility to jurists and, as Passmore has indicated, to moral and legal philosophers ('see particularly the distinction between "description" and "ascription" in H. L. Hart'). He requires that we study *actual* languages, *not* ideal ones' and one can all too readily envisage the contempt with which he would dismiss Santayana's assertion that 'mind is incorrigibly poetical' in its transmutations of material facts and practical exigencies into 'many-coloured ideas'. If such etiolations were the truth of poetry one would be bound to favour Austin's prosaic method, 'taking the sentences *one at a time*, thoroughly settling the sense (or hash) of each before proceeding to the next one'. Performative utterances that are '*in a peculiar way* hollow or void', that cannot be said to enact, contrive to avoid taking the rap for their own claims. In this respect the *MHRA Style Book* is basically Austinian:

> Avoid the practice of using quotation marks as an oblique excuse for a loose, slang, or imprecise (and possibly inaccurate) word or phrase. Quotation marks should normally be reserved to indicate direct quotation from other writers.

One must avoid casuistry; and quotation marks, except as an indication of direct quotation, are a casuistically ideal language masquerading as an actual one. On consideration, however, this proscriptive view appears too simplistically exclusive. In Pound's use of quotation marks in *Homage to Sextus Propertius* and *Hugh Selwyn Mauberley* ('"Wherefrom father Ennius, sitting before I came, hath drunk"'; 'The "age demanded" chiefly a mould in plaster') the effect is not that of avoiding the rap but rather of recording the rapping noise made by those things which the world throws at us in the form of prejudice and opinion, 'egocentric naïveties' and 'obtuse assurance'. The *MHRA Style Book* inculcates one excellent kind of fidelity, to the patient accumulation of data about actual usage, but it also suggests, by tacit implication and even while urging upon us a strict acceptance of personal responsibility, that not only the data contained in 'direct quotation from other writers' but also the medium of language itself are more transparent, more innocent, than we have any right to expect. It does not forewarn us that, when we quote, we are necessarily engrafting, together with scrupulously accurate formulations, much loose, slang, or imprecise matter and many 'compacted doctrines'. Nor does it indicate how these obliquities are to be registered, except by some ideal scientific neutrality of tone. Yet when Austin states that 'philosophers often seem to think that they can just "assign" any meaning whatever to any word', his own use of quotation marks indicates the limitations of the basically utilitarian approach to 'the innumerable and unforeseeable demands of the world upon language' and concedes that a touch of miming may be a proper part of the process of assessment. When G. H. Lewes admired Jane Austen for, among other things, her lack of sentiment, Charlotte Brontë retorted 'you scornfully enclose the word ["sentiment"] in inverted commas'. Here the demands of the world upon language are being mutually though incompatibly registered. And when we read how the First World War soldier, in his disillusionment, 'turned and picked up some of the fine phrases which had stirred his heart' so that '"Remember Belgium!" was heard with ironic and bitter intonations in the muddy wastes of the Salient' we encounter the same phenomenon in a more extreme manifestation. As Coleridge said, 'our chains rattle, even while we are complaining of them'. 'Inverted commas' are a way of bringing pressure to bear and are also a form of 'ironic and bitter' intonation acknowledging that pressure is being brought. They have a satiric function, can be used as tweezers lifting a commonplace term out of its format of habitual connection. That which the *MHRA Style Book* sees as 'oblique excuse' may in fact be direct or oblique rebuke. 'Our word is our bond' is an exemplary premiss, but if we take it in the positivist sense alone we take only part of its mass and weight. 'Our chains rattle . . .' is the inescapable corollary for any writer who takes up Austin's stringent sentiment. At the same time, the necessary consequence of admiring Austin's logic would seem to be that there is nothing to take up. 'If said by an actor on the stage, or if introduced in a poem, or spoken in soliloquy' our ideal terms can have no actual consequences. Does it not necessarily follow, therefore, that the audacity of Pound's 'all values ultimately come from our judicial sentences' is achieved at the expense of a missed connection, precisely the one which Austin picks up in his caveat about 'the final arbiter of . . . the phenomena'? In claiming 'ultimate' sanction Pound is merely 'surveying the invisible depths of ethical space', a practice which, in Austin's view, permits a variety of moral 'welshing'. Poetry has no 'phenomena', in Austin's sense, only 'noumena' or perhaps 'nous', that 'act of mind' which is also the 'sense or meaning' of its own statements.

There would seem to be, as we have said, much cherishing prescience as well as shrewdness in Sidney's contention that the poet 'nothing affirmeth, and therefore never lieth'. It appears at least to offer a compromise whereby the fiction can be given its proper status precisely because it does not claim that it could be 'seriously performed': 'it is that faining notable images of vertues, vices, or what els . . . which must be the right describing note to know a Poet by'. The thought occurs to us that the poet is here being generously propositioned. If he will accept that his art is a miniature emblem or analogy of res publica rather than a bit of real matter lodged in the body politic there is much scope for the exercise of serious and refined example. Sidney certainly invites a high degree of imaginative responsibility, of the kind and order that is manifested in Sonnet LXVIII of Spenser's *Amoretti*:

> This ioyous day, deare Lord, with ioy begin,
>> and grant that we for whom thou diddest dye
>> being with thy deare blood clene washt from sin,
>> may liue for euer in felicity.
> And that thy loue we weighing worthily,
>> may likewise loue thee for the same againe:
>> and for thy sake that all lyke deare didst buy,
>> with loue may one another entertayne.
> So let vs loue, deare loue, lyke as we ought,
>> loue is the lesson which the Lord vs taught.

Within limits, this exercise of semantic judgment is exemplary. Christ's blood is 'deare' because it is precious; it is precious because it was the price of man's ransom and it cost Christ dear in terms of dire suffering. The tenderness between betrothed lovers ('deare loue'), the fealty of servant to master ('deare Lord'), are only possible because of the hard, grievous sacrifice of Christ. What happens here is more solemn than a play of wit; it is a form of troth-plight between denotation and connotation. Spenser's text is also wedded to the text of 1 Corinthians: 'ye are dearely bought'. The 'felicissime audax', the 'curiosa felicitas', of pagan rhetoric are reborn and baptized in Spenser's 'felicity'. But it is a felicitousness which knows its place and which achieves its directness by not being presumptuous, by respecting priority and status. The intelligence may be consummate but the consummation is elsewhere, as the stance of the final couplet suggests. 'So let vs loue, deare loue, lyke as we ought' is already leaving the poem behind; the 'lesson' to which the poet directs us is, as the Austinians would say, 'extra-linguistic'. It must also be said that, when we read *Amoretti* in its entirety, the felicity of LXVIII appears uncharacteristic of the sequence, which terminates in quarrel and separation. The Pauline cadences of this sonnet are unheralded by the Petrarchan cadences of LXVII and leave no posterity in the twenty-one sonnets which succeed, for these fluctuate between clichés of loss, absence, mortality, on the one hand, and, on the other, appeals to the superior virtues of 'contemplation', of 'beholding the Idaea playne'. It is not possible to determine precisely in what proportion this sense of relapse is due to artistic impatience, the 'haste and botching' at the time of publication or to Spenser's desire to embody a Sidneian awareness of the gulf between 'erected wit' that 'maketh us know what perfectio[n] is' and 'infected wil' that 'keepeth us fro[m] reaching unto it'.

Theodore Spencer observed that Sidney 'extolled poetry as greater than either of its rivals, philosophy and history, for it combined the wisdom of the one with the concrete examples of the other'; but this is an observation that calls for some modification. Sidney writes on occasion as though recognizing that the poet is in reality forced to compound his '*Idea*' of wisdom with the 'particular truth of things', that, like the historian, he is 'captived to the trueth of a foolish world'. The likeness of these phrases exacerbates the degree of their unlikeness. Sidney, even while composing a 'Defence', is resisting two possible defensive postures: one, that the coercive power of the 'particular truth of things' makes all 'truth' merely contingent and relative; the other, that the status of the 'idea' provides the poet with an Archimedean ec-stasis, a station above and beyond the world's gravity and folly, a place of serenely measured hypotheses. Fulke Greville offers a simpler gloss on his friend's beautifully guarded but honest involvements and returns: 'For my owne part, I found my creeping Genius more fixed upon the Images of Life, than the Images of Wit'. 'Creeping Genius' has a blunt 'foolishness' which Sidney eschews in his treatise though not in such poems as Sonnet 71 of *Astrophil and Stella* where 'erected wit', pursuing the 'idea' of a Platonic Stella, is, by the truth of a foolish world, moved by carnal impulses: 'But, ah, desire still cries: Give me some food'. 'Erected wit' has, after all, inspired merely 'an erection of those engendring parts'. This ironic realization comes as a perfectly calculated surprise and one has to distinguish between such effects, where the poet is winningly in command of his own perplexity and weakness, and passages within which, or between which, there lies an area of real perplexity, 'dark and disputed matter', problems which the writer 'unconsciously' raises.

We return here upon an 'ineluctable' problem: that those for whom writing is like 'bearing a part in the conversation' must regard with incomprehension those for whom it is 'blindness' and 'perplexity' and that those for whom 'composition' is a struggle with dark and disputed matter will inevitably dismiss as mere worldliness the ability to push on pragmatically with the matter in hand. The theologian D. M. MacKinnon is characteristically scrupulous in his observation that 'rich possibilities' may depend upon 'quick practical perception' and justly chastening when he notes that 'the most intimate self-interrogation' can quickly assume 'the highly questionable status of the subjective and capricious'. At the same time, T. H. Green's 'there is that in us which is the negation of each of our acts, yet relative to each of them, and making them what they are' takes account of the highly questionable nature of the 'objective status', if this is seen as some Archimedean viewing-platform for 'naked undisguised ideas' 'stripped of words'. To restrict MacKinnon's 'quick practical perception' to this kind of functional immediacy would be to misconstrue him as gravely as one would misrepresent Austin by implying that 'transparent' means anything like 'rudimentary' or is even directly equatable with Sprat's 'primitive purity and shortness'.

In referring to ineluctable problems, therefore, we would do well to consider Austin's warning that 'we can know the facts and yet look at them mistakenly or perversely' and, in that light, to take up the question whether, in arguments of this kind, certain transcendental explanations are preferred because they appeal merely to temperamental proclivities which incline one to favour either a comic or a melodramatic view of such matters. The comedians, according to this theory, would favour notions of 'felicity', the 'providence of wit'; the melodramatists, searching the same facts, would apply a more sombre tincture: 'infected wil', 'concupiscence of witt', 'anarchy of witt'. We could emblematize the possibilities in the following way. Cudworth, in his *Sermon Preached Before the House of Commons*, 1647, warns against 'shaping . . . out' God 'according to the Model of our selves, when we make him nothing but a *blind, dark, impetuous Self will*, running through the world; such as we our selves are furiously acted with . . .' We feel at once how the density, the weight, of this statement manifests itself in the very concept to which, Cudworth says, we should not give weight. To maintain that the italicized words are the most ponderable is to suggest no more than that Cudworth, while declaring that God is not the image of our solipsism, mimics our solipsistic furor. To take '*blind, dark, impetuous Self will*', out of context, as motto or epigraph (and the printer's italics offer us that temptation) is melodramatic and perversely stimulating. This sermon anticipates by a few years Vaughan's 'But living where the Sun/Doth all things wake, and where all mix and tyre/Themselves and others, I consent and run/To ev'ry myre . . .' It is true that Vaughan is speaking of sin, not language; but one might insist, subjectively and capriciously, that precisely because one can refer so readily and properly to 'the run of the verse' or the run of a sentence, the writer's vocation is the one which 'consents' most readily, which 'runs' most directly, to its own confusion. It is, in a peculiar way, the most oxymoronic art: its very making is its undoing. In a poet's involvement with language, above all, there is, one would darkly and impetuously claim, an element of helplessness, of being at the mercy of accidents, the prey of one's own presumptuous energy. The salutary pithiness of Thomas Hobbes springs from his unlamenting, sardonic observation of this affliction, in the brusque 'for if he would not have his words so be understood, he should not have let them runne'. And the comedians, from Nashe to Beckett, could also insist that the undoing of language is, as often as not, the making of it. J. L. Austin is himself a writer in the comic tradition. His radical 'doctrine of the *Infelicities*' could not be more happily expressed.

'Comic tradition' is, however, an infected term. The system within which Austin exercises his discriminations is not quite a comedy of manners and not quite a line of wit. When a man named Austin entitles his lectures 'Sense and Sensibilia' he is simultaneously being precise and asserting something; he is accepting the gift, the aptness of the thing given, and he is displaying his own 'gift', his aptitude for making the most of the *donnée*, in a pleasing way, to himself and to us, though the pleasure is of a minor kind. We are tickled and suspect that our feeling coincides exactly with the author's. G. J. Warnock remarks, with unnecessary diffidence, that 'not everyone, I imagine, found Austin's jokes – including the really silly ones – as funny as I did'. In *How To Do Things With Words* the conclusions are embodied in a final series of comparative word-lists. Nonetheless, the advocate of patient accumulation also 'liked authority' and enjoyed a keen 'sense of his own position'. He minutely and accurately discriminates the senses of English words and phrases, and instructs us how we may 'learn from the distinctions encapsulated in their ordinary uses'. To drive with attention may not be the same as to drive 'with care'; 'clumsily he trod on the snail' is to be distinguished from 'he trod on the snail clumsily'; 'without effect' does not (in a particular circumstance) mean 'without consequences, results, effects'. He is at the same time a polemicist of distinctive irony. A reference to the jurists' 'timorous fiction, that a statement of "the law" is a statement of fact' contains a crucial distinction but is distinctly different from the 'patient accumulation of data'. There is a pleasurable impatience in 'timorous'. 'Clumsily he trod on the snail . . .' is at once a discrimination of meaning and a nuance about the heavy-footedness 'out there'. One might feel, however, that it is entirely just for Austin's method to hover between manners and wit since, as we have observed, one of his epistemological cruces is the necessity to meet the innumerable and unforeseeable demands of the world upon language. The world and the word are equipoised and Austin's sense of occasion is as strong as his sense of semantic propriety. The disposition of his Harvard seminars ('In these discussions the physical and dialectical centre of gravity located itself, predictably, in the person of Austin') can be taken as a paradigm of language's connection with the 'world'. But how is the 'world' to be reckoned in this instance? Primarily it is the 'magic circle' of the philosophers, described in Berlin's memoir, with a larger circumference than David Hume's but held together by similar ideals and expectations. It is a 'World', the attention of which one seeks to attract by 'Elegance and Neatness' of address, the indifference of which condemns one to calamitous 'Obscurity'. Yet, at the same time, that which the seminar concentrates upon is nothing less than the 'world', epistemologically and ethically conceived, ranging from the implications of Sidney's 'most excelle[n]t resting place for al worldly learning' to those of Hopkins's ambiguous, ungraspable 'world-wielding' force ('something that makes, builds up and breeds . . . something that unmakes or pulls to pieces . . .'). The world of which Austin takes account in his quips and parables is basically cheerful, hedonistic, preoccupied with business, professional conduct and games-playing but it is also shot through with anger, infelicity, blank incomprehension:

> The man himself, such is the overriding power of the pattern, will sometimes accept corrections from outsiders about his own emotions, i.e. about the correct description of them. He may be got to agree that he was not really angry so much as, rather, indignant or jealous, and even that he was not in pain, but only fancied he was.

If we take the full force of 'overriding', the 'pattern' will appear to be coercive, proscriptive. However, 'working the dictionary' as Austin prescribes, we note that historical usage infects the word with a variety of 'considerations'. It may be 'unconstitutional' to override but constitutions themselves have the power to do so. There is a worldly ambivalence in Austin too: an Occam-like paring down, a brusque rebuttal of '*insouciant* latitude' on the one hand; on the other a tolerance, though not necessarily an acceptance, of routine. The world wags vigorously in his work and he is also a bit of a wag. The man who 'only fancied' he was in pain belongs to the same comic order as the faith-healer of Deal. Austin, we are told, 'could not bear histrionics', 'gaudy pronouncements', 'escape into metaphor or rhetoric or jargon or metaphysical fantasy'. It is itself a philosophical irony that a mind which strove for accuracy of definition while registering most acutely the quotidian duplicities, which sought 'decent and comely order' as fervently as did the authors of the antique tropes, felt free to regard poetry as one of the non-serious '*parasitic*' '*etiolations* of language', as a kind of 'joking'.

One's perplexity in this matter is compounded by the recognition that, notwithstanding the force of his slighting asides, there is a 'poetic' quality in Austin's writing at its best, as, for example, in the finesse with which, in successive paragraphs, he moves from the ironic negative 'spiritual assumption of a spiritual shackle' to the 'plain saying that *our word is our bond*'. In this sequence the proscriptive sense of 'arbitrary constraint' and the affirmative sense of 'covenant' are firmly and reciprocally adjusted. By 'poetic' in this context one means no more than that the richness and subtlety of 'ordinary language' have been amply realized, not simply run into or stumbled over by someone who is 'flurried' or 'anxious to get off' as Austin would say; but it remains a quality to be noted, and one which he himself might have noted more. At one point he justifies the use of 'jargon' to explain varieties of behaviour 'which have not been . . . hallowed by ordinary language'; at another he writes that 'superstition and error and fantasy of all kinds do become incorporated in ordinary language'. Here again we may note that the way in which 'hallowed' and 'superstition' jar slightly against each other's grain constitutes Austin's way of setting a 'plot'. Though his prevailing thesis is that our function is to 'clear up . . . usage', 'arrive at . . . meanings', a counterpointed idea is that 'the word snares us' and that we may 'cheerfully subscribe to, or have the grace to be torn between, simply disparate ideals'. One can see precisely why a parenthetical clause is as far as he can go towards desperation, since any further move in this direction would be deplorably towards 'the old Berkeleian, Kantian ontology of the "sensible manifold"'. So far, we may say, Austin's ethical toughness is nothing less than admirable; and, so far, his feel for the tension between 'cheerfully', 'grace', and 'torn' has the fineness of poetic tact. But only so far. When Austin's cheerfulness veers towards contempt he seems to care less: 'a_2["clumsily he trod on the snail"] might be a poetic inversion for

148

b₂ ["he trod on the snail clumsily"]'. 'Poetic' exists to cause offence. It is a parody, the limits of which are not precisely drawn and which are therefore variable according to Austin's particular whim. Iris Murdoch suggests that 'in a way' he '"saves" . . . the old impersonal atom-world of Hume and Russell' 'by substituting an impersonal language-world'. This describes the intention or the effect fairly enough; but his own language is not really 'impersonal' at all. The one 'infection' that he fails to realize (in the semantic fullness of that word) is the infection of wit by manners; the one 'superstition' he does not wholly eradicate is the mystique of his own superior intelligence.

Things seem, and perhaps are, tougher now than they were, or appeared to be, when R. L. Nettleship wrote in 'Language and its Function in Knowledge': 'the only proof that language really *is* communication, and that there *is* a mutual understanding . . . is that we act on the belief that there is, and that this belief is justified by the results'. Taken out of context, however, this sentence makes Nettleship appear more of an unperturbed utilitarian than in fact he is. He is closer to the Coleridgean 'drama of reason' than that:

> Words are symbols made by us, but we come to look upon them as mysterious agencies, under whose power we are . . . If we all tend to become the 'victims of words,' the corollary is that we should mean something by our words, and know what we mean. We cannot get over the difficulty by blaming language or declining to use it.

The impetus of Nettleship's argument seems to require him to take the Coleridgean sense of words as 'LIVING POWERS' and to re-express it as a negative fallacy; but he does not quite do that, even though 'mysterious' is used critically not mystically. By working with the 'we' and 'if' clauses the critic chooses to remain 'within the process'. If we follow T. H. Green's argument, as put forward in his paper 'The Philosophy of Aristotle', to place ourselves 'outside the process by which our knowledge is developed' is to conceive of an untenable 'ecstasy', whereas to recognize our being within the process is to accept our true condition. It must nonetheless be accepted that, in the work of the so-called British Idealists, there remains a residuum of words which, while appearing weighty and stubborn, are in fact fluid and amenable. In this respect the work of Austin and his followers, who are suspicious of 'ideal' languages and who desiderate the 'transparency' of description, may be seen to be a radical departure and to attempt a radical criticism. Green's use of the Kantian 'manifold' is in part a technical gloss and in part an impressionistic tincture. Much the same could be said, however, of certain remarks by Austin which fall into a semantic hiatus between two individually distinct tones: one of empirical description, the other of 'malicious pleasure'. With 'J. L. Austin/*Sense and Sensibilia*' we know precisely where we are; with 'if the poet says "Go and catch a falling star" or whatever it may be, he doesn't seriously issue an order' we suspect raillery.

Coleridge observed that 'to a youth led from his first boyhood to investigate the meaning of every word and the reason of its choice and position, Logic presents itself as an old acquaintance under new names'. In post-Coleridgean chronology these 'names' include poetry and 'linguistic phenomenology'. By 'poetry' one means, in the context of this discussion, 'troped' utterance, 'ontological manoeuvre', sometimes 'desperate', sometimes felicitous, occasionally marked by that strain of desperate felicity which R. P. Blackmur noted as a characteristic of 'the classical mode in the arts' in modern times. It is 'as if order *required* distress'. 'Required' is, in the circumstances, well-judged, hovering as it does between questions of temperament and technique. Kenneth Burke has described 'workmanship' as 'a trait in which the ethical and the esthetic are one'. The appeal of this statement rests in the conviction that a formula exists for achieving a consummation of technique which simultaneously 'satisfies the desire of a moral agent' and, in so doing, resolves the 'old difficulty', as it has been called, 'of conceiving . . . *an activity with end attained*'. Both the New Criticism and Austinian verbal analysis are in this respect scions of that passage from *Biographia Literaria* in which Coleridge aligns 'verbal precision' and 'mental accuracy', at once prescribing 'modes of intellectual energy' and proscribing 'fanaticism', 'indistinct watch-words'. There is here no division to be inferred between 'energy' and 'precision'. Austin's objection to this might be that Coleridge's own terms constitute nothing more than indistinct watch-words, metaphysical fantasy, pretence. '*Praetendere* in Latin', he writes, 'never strays far from the literal meaning of holding or stretching one thing in front of another in order to protect or conceal or disguise it: even in such a figurative use as that in Ovid's "praetendens culpae splendida verba tuae", the words are still a façade to hide the crime'. Austin's reference is to the *Remedia Amoris*, lines 239–40, 'Nec te Lar patrius, sed amor revocabit amicae . . .', and 'crime' is, in fact, a shade too portentous a rendering of the 'weakness' involved in letting oneself be lured back by sexual desire and in disguising that motive in brave words about home and country. There is indeed a range of ironic implication in Ovid's use of 'culpa' and 'splendida verba' ('the grand name without the grand thing'), his realization of the manner in which 'holding forth' can be a way of withholding, which Austin chooses not to point out, either because he takes it to be self-evident or because he is excluding from consideration anything to do with 'that rather woolly word "imply"'. A poem, in his opinion, is not 'issued in ordinary circumstances'. This makes sense. What also makes sense is Ransom's 'the density or connotativeness of poetic language reflects the world's density', an observation which Austin might also dismiss as metaphysical fantasy. Those invocations at the heart of Hopkins's 'The Wreck of the Deutschland' – 'O Deutschland, double a desperate name!', 'Double-naturèd name,' 'O unteachably after evil, but uttering truth' – would be, from that angle of vision, incorrigibly self-stultifying. Notwithstanding such logical objections it must be affirmed that it is at such points, or nodes, where 'stultification' might seem the most reasonable verdict, that poetry encounters its own possibilities. The word 'blind' upon which Wordsworth homes in so often is itself a compounding of blankness and intuition: at one time 'Dim sadness, and blind thoughts I knew not nor could name' at another 'A pleasurable feeling of blind love'. Those other Wordsworthian key-words 'gleam' and 'perplexity' gather their reciprocating force from this blank recognition: from the sense that without the perplexity there would be no gleam and that the 'blindness' embodies them both. When Hopkins writes 'dogged in den' he does not mean what he means by 'dear and dogged man'. Yet his early fathoming of words as 'heavy bodies' bears most fully on this matter. He is drawn down to a double nature within the etymological stratum, where *dǫgd* (hounded) and *dǫ·gèd* (tenacious) lie like shards or bones of 'most recondite and difficult' matter within the simple hereditary accruals of the vernacular. It is 'man's malice' that is 'dogged in den', a sense at once opposed to the 'dear and dogged' and yet inextricably tied to it. *Dǫ·gèd* itself once meant 'currish', 'sullenly obstinate'; that it came gradually to signify something more 'dear' than this is due among other things to Dr Johnson's decision, in 1779, to refer to 'a dogged veracity'.

We are therefore driven to look again at Nettleship's unwillingness to accept the idea of words as 'mysterious agencies, under whose power we are' and to insist, despite having sympathy with this admirable form of common sense, that there *is* something 'mysterious', some 'dark and disputed matter' implicated in the nature of language itself. But the mystery is nothing more nor less than 'ordinary circumstances', 'habitudes and institutions', 'cultivated opinion', 'traditional pieties and naïve beliefs', what Locke termed 'the audible discourse of the company' and Austin designated as 'the conduct of meetings and business'. 'Our word is our bond' (shackle, arbitrary constraint, closure of possibility) is correlative to 'our word is our bond' (reciprocity, covenant, fiduciary symbol). 'Mastery' is as much as is not servitude. The 'ineluctable problem' is therefore, as might reasonably be expected, elementary and empirical. To encounter it, however, is like encountering a blank in one's own thinking where '*bona fide* perplexity' is hardly distinguishable from obtuseness, and instinctive flinching from disingenuous evasion. Donne, for example, freely invents 'paradoxes and problems' but he also has problems that are not paradoxes, that cannot be 'impudently' troped but must be rawly acknowledged. 'I did best when I had least truth for my subjects' is aware that it calls 'best' into question; though this observation is more oblique than those statements, in *Devotions upon Emergent Occasions*, which bind the erected wit into the mortal sickness of man's engendering parts ('to lay him sicke in his owne bed of wantonnesse') or which, still more blankly, set the foundations of our helplessness even deeper than the region of the 'infected wil':

> Or if these occasions of this selfe-destruction, had any contribution from our owne *Wils* . . . we might divide the rebuke and chide our selves as much as them . . . But what have I done, either to *breed*, or to *breath* these *vapors*? They tell me it is my *Melancholy*; Did I infuse, did I drinke in *Melancholly* into my selfe? It is my *thoughtfulnesse*; was I not made to *thinke*? It is my *study*; doth not my *Calling* call for that?

If we were to thrash this matter out, either with him or against him, we could argue that Donne, while conceiving of a passivity which he strives to separate from malignant intention, precludes, in that very conceiving and striving, the completeness of the distinction. He argues that in '*Fevers* upon wilful distempers of drinke, and surfets, *Consumptions* upon intemperances, and licentiousness' and the like 'our selves are in the plot, and wee are not onely *passive*, but *active* too, to our owne destruction'. This is not, we may say, like having one's proper 'calling', one's ordained faculties, turned into the very accessories of destruction. Some thirty years before Donne's *Devotions* were printed, the public stage had exhibited Marlowe's Faustus wilfully and perversely confounding licentiousness with study. There remained sound reasons for Donne's desire to draw a moral distinction between them. Even so, one might still quibble with his apparent inference that the metaphor 'our selves are in the plot . . .' in some way ceases to apply when we abandon 'distempers', 'intemperances, and licentiousness' in favour of 'thoughtfulnesse' and 'study'. It could be contended that what T. H. Green calls being 'within the process' and Donne calls being 'in the plot' is the peculiar nature and burden of that activity we are accustomed to call 'thinking experience'. It is within the process of such experience that we are not only active but passive too, exhibiting the symptom at the very moment that we diagnose the condition. But this is something more openly empirical, less amenable to the simple revenge cliché of the libertine sinner laid sick in the bed of his own 'wantonnesse'. It is more the recognition that our '*Noble* parts . . . for all their priviledges . . . are not priviledged from our *misery*'. Donne has not, after all, missed or evaded the point. Nygren says that 'in Augustine, the sinful soul is "*bent down*" *to earth*; in Luther, it is "*bent upon itself*"'. Whether we ourselves cling to the Augustinian or to the Lutheran emphasis our conclusion must be that the language a writer uses and the writer who uses the language are inextricably involved and implicated. If Hopkins's words are 'heavy bodies' they are 'bent down to earth'; if the creative spirit is necessarily 'bent upon itself', then its deepest intuitions are ineluctably compounded with its most inveterate stubbornness and incapacity. This is the perplexity upon which Donne briefly but searchingly touches when he alludes to our being in some sense vocationally compromised, condemned by the very nature of our 'calling'.

To the reasonable retort that one is 'vocationally compromised' only by being temperamentally predisposed to view matters in this dark light, by an inclination to the melodramatic, one may reasonably concede that the 'hazardous course', the 'partial surrender', the 'adjustment for practical purposes', are unavoidable and unarguable ('our word is our bond') but that one is not bound by anything other than temperament and indoctrination to the manner in which the exigencies are viewed. But even here one might be conceding too much, simply to oblige. There have always been rich opportunities for confusing temperament with technique and it may be inequitable to wish the confusion upon some chosen author. Dryden, for instance, who in one place celebrates 'the providence of wit', in another deplores its 'anarchy'. It would be unwise to suppose some violent change of mood on his part, but reasonable to assume his customary pragmatic acceptance of the need for an altered mode. 'Infections' are 'ordinary circumstances' and the dyer's hand, steeped in etymology if nothing else, is, by that commonplace craftsmanlike immersion, an infected hand. One can say with Augustine 'curva voluntas' or with Peirce 'actuality is something *brute*. There is no reason in it. I instance putting your shoulder against a door and trying to force it open against an unseen, silent, and unknown resistance'. Either way this is the gravamen of the matter, but the matter may of course be an occasion for levity.

Austin himself has remarked that 'we must consider the total situation in which the utterance is issued – the total speech-act'. If one accepts the paradigm of the warning notice which in 'saying that customers are warned *is* warning the customers' one is not finally annulling the extra-linguistic sanction of convention. As Warnock notes, such a warning contains, either explicitly or implicitly, the word 'hereby'. '"Hereby" is an indication that the utterance *itself* is doing the job that it says is done'. This is, by general consent, one of the most compelling instances in which to *say* something is to *do* something; and yet the inescapable correlative seems to be that 'hereby' can do what it says only because there exists some idea of sanction (real or fictional) to back it up. Modern poetry, we may suggest, yearns for this sense of identity between saying and doing – 'all values ultimately come from our judicial sentences' – but to Pound's embarrassment and ours it discovers itself to possess no equivalent for 'hereby'. It does not seem possible to call into court, on Poetry's behalf, those other 'necessary circumstances' which would establish its claim to be a juridical agent, except in a grotesquely negative form. For anyone predisposed to regard Pound's statement as mere 'holding forth', 'splendida verba' cloaking self-delusion, inane 'culpa', the Faustian catastrophe would have been enacted outside the Washington courtroom in 1945: 'When I asked him whether he wanted to stand mute or would prefer to enter a plea, he was unable to answer me. His mouth opened once or twice as if to speak, but no words came out'. But the silence is here too eloquent a rebuke to the 'inviolable voice'. If, in Pound's humiliation, we have an exemplary instance of 'idea' brought face to face with 'the particular truth of things', we have equally a parable of the due process of law enmeshed in various 'self-stultifying procedures' ('A. . . . he comes back to them [i.e. 'fixed' ideas] all the time, but none of them is clear. It is all vague . . . Q. You mean vague to the examiner? A. Yes, of course'). MacKinnon is right: 'rich possibilities' do depend upon 'quick practical perception' and the distinctions between the absolutes and partialities of justice are a matter of quick practical perception too. 'All four experts . . . attested a prevailing "grandiosity" in the poet, which they said indicated his abnormal mental state.' It may be grandiose or 'delusional' to claim that one possesses 'the key to the peace of the world through the writings of Confucius' since so unworldly a sense of the world must be, by that definition, self-stultifying, oxymoronic, impotent enactment, 'hollow or void' as though introduced in a poem or spoken in soliloquy. At the same time, the 'peculiar legal paradox' that, as a result of the court's verdict, 'Pound found himself, in effect, under a sentence of life imprisonment despite the fact that he was innocent in the eyes of the law' is in itself more oxymoronic than paradoxical.

This is not said to let Pound off the hook, or to spare him the rap. Cornell's suggestion that he 'was imprisoned despite the fact that he is one of the great literary figures of our time, perhaps in part because of that fact' is 'incorrigibly poetical' in Santayana's dubious sense. Pound, in his *Paris Review* interview, said that 'there is the struggle not to sign on the dotted line for the opposition'. His own idiom implicates itself in 'the conduct of meetings and business'. Our chains rattle, even while we are complaining of them. The moral offence of his cruel and vulgar anti-semitism does not call into question the integrity of his struggle; neither does the integrity of the struggle absolve him of responsibility for the vulgar cruelty. The essential culpability of his wartime broadcasts was not their eruption into 'that stupid suburban prejudice', as he self-indulgently called it, but their 'insufficient desperation', as Richard Reid has acutely observed. The more important word here is 'insufficient'. 'Saeva indignatio' is no guarantee of verdictive accuracy, or even of perception; and it is lack of attention, or 'care', which brings Pound to the point of 'signing on the dotted line' for the rulers of the darkness of this world – not in spite of, but through, the mundane struggle, the 'being bound' to push on with the matter in hand, no matter what, where the matter is the 'heavy bodies', the 'solid entities', the 'compacted doctrines'.

T. H. Green argued that, in revolutionary Jacobinism, Hume's 'philosophy of feeling . . . had by a necessary process recoiled upon itself'. He had in mind that inert compound of 'wilfulness' and acquiescence to 'necessity' which he regarded as the negative side of hedonistic materialism. 'Recoil' upon the self should therefore, I believe, be distinguished from that 'return upon the self' which may be defined as the transformation of mere reflex into an 'act of attention', a 'disinterested concentration of purpose' upon one's own preconceived notions, prejudices, self-contradictions and errors. Viewed in this light Pound's ultimate remorse ('I guess I was off base all along'), though not baseless, was unbalanced; a recoil rather than a return. In a note scribbled to his defence lawyer from St Elizabeths Hospital he wrote:

> enormous work/to be/done./& no driving/force/& everyone's/inexactitude/very/fatiguing

'Everyone's inexactitude' (if, against the run of that bias, we include his own) comes closer to the heart of the matter than 'off base all along'. 'All values ultimately come from our judicial sentences' sounds magisterially Shelleyan but in fact does less than justice to Pound's grasp of *logopoeia*. Christopher Ricks's redefinition, 'all values ultimately go into our judicial sentences', stands the perception in its rightful place, at the receiving end of 'the innumerable and unforeseeable demands of the world upon language', rather than in a position of vatic privilege. Since Pound places such emphasis upon definition ('The poet's job is to *define* and yet again define till the detail of surface is in accord with the root in justice') it is not unjust to take him up precisely on this point, on this disjunction of the aesthetic and the ethical. Coleridge observed that 'the cultivation of the judgement is a positive command of the moral law'. In such pronouncements as 'all values ultimately come . . .', 'positive command' detaches itself from 'cultivation of the judgement'. The resultant tone strikes one as being nothing more than the assertion of status; and 'I guess I was off base all along' becomes complicitously egocentric.

The usefulness of Ricks's redefinition is that it both assuages and aggravates one's awareness of the obligation to judge definitively those matters which one is arbitrarily stuck with. To take a stand upon such questions is one thing; mere vatic 'status' is another. 'Fact is richer than diction' as Austin says, but there seems no just cause to infer from this a parasitic role for poetic statement. Of *Homage to Sextus Propertius* Pound wrote 'I certainly omitted no means of definition that I saw open to me'. The final line of the sequence:

And now Propertius of Cynthia, taking his stand among these

conflates two lines of Propertius' Latin: 'Cynthia quin etiam versu laudata Properti / hos inter si me ponere Fama volet'. The passive ('hos inter si me ponere') becomes the active 'taking his stand'. The major implications of the change certainly involve a claim to status, to be 'among' the true poets. The minor nuances include a hint of proud vulnerability (having to stand up for oneself in such a world) and the fleeting glimpse of a cocksure literary barrow-boy ('Near Q. H. Flaccus' book-stall'). The status fought for, and accomplished, within the comedy and melodrama of this sequence, is, therefore, that of standing by one's words in a variety of tricky situations and is a different matter from abstractedly surveying the depths of ethical space ('all values ultimately come . . .'). The tone of that last line is stubborn, jaunty, and yet elegiacally aware of the tragic farce of being bound to maintain standards against such odds. Each of these tones is appropriate in its impropriety, given the 'imbecility' against which it is pitched. It is true that 'imbecility' is Pound's prejudice, but *Homage* is aware of prejudice and other coarsenesses, as it allows that 'fact is richer than diction' in the very process of evolving a means of defining its response to matters of such perplexity.

Contemporary aesthetics, when they are being ethical (as in the exegetical work of Nathan A. Scott), seek 'exemplary instances of the literary imagination supervising its project'. This leads, all too often, to a style of appraisal which is little better than a series of exclamation-marks: 'extraordinary adroitness', 'wonderfully nuanced and deeply affecting', 'astounding range of . . . vocabulary', a form of that Orphic afflatus which the critic elsewhere views with some unease. Yet we are required to take inflationary adjectives as a true evocation of the poet's 'redeeming work' in face of 'the world's absurdity'. There is therefore an inherent irony in meditating, as Scott does, upon 'absolute closures of possibility' in terms so effusive and grandiose. Such 'closures' are consistently seen, by practitioners of this mode, as thematic data, as matter for discursive explication; but it is the modish style itself which, with its simplistic affirmations, most drastically forecloses on other possibilities. The vigorous pursuit of misery, degradation and guilt thus appears in a peculiar way ebulliently heartless and obtuse. In itself, however, 'absolute closures of possibility' is a phrase worth pondering. It merits being taken, not in Scott's style of defining a superficial sense of the 'Absurd', but as something more closely approximate to that 'foundational' thinking, that grasp of 'primal reality', which he elsewhere wishes upon his hypothetical figure of 'the poet'. The approximation would be closer to Greville's 'creeping Genius' or to Sidney's historian who 'wanting the precept, is so tied, not to what should be, but to what is' than to Scott's Heideggerean adept; but such humiliation would be Scott's 'foundational thinking' with a vengeance.

To return upon ourselves. When Hopkins writes of 'dark and disputed matter' he is not, in the first instance, thinking of linguistic manifolds. He is meditating primarily upon the nature of voluntary and involuntary acts, the 'active' and 'passive' will, the 'free' and the 'constrained'; upon the distinctions between these which must be drawn and observed, and upon the numerous cross-thwartings which make the maintaining of such distinctions an arduous and perilous task. There is an essential 'freedom of pitch' (which is when 'I instress my will to so-and-so') and there is 'accident[al]' 'freedom of field'. If, however, language as medium is a prime manifestation of 'freedom of field' and the right-keeping of will manifests 'freedom of pitch', Hopkins's theological crux is necessarily a linguistic crux. The abrupt and ugly phrasings ('*the doing* be, *the doing* choose, *the doing* so-and-so . . .') excessively, even absurdly, concentrate the sense of pitch. It is a consciousness which accepts that the determining of grace necessitates at times a graceless articulation: we are reminded that Hopkins was, albeit briefly, a pupil of T. H. Green. But he was also a student of Plato. As a number of his poems reveal, he would regard it as more desirable that grace should be gracefully embodied and declared. Felicity, always to be desired, is met with in unlikely places, even in a 'hantle of howlers' scattered across the broad expanse of freedom of field: 'Caesar is proud and pompeous', 'The Wife of Bath was a real woman's man', 'Swift's misagony'. As Edwin Morgan remarks, these 'are hardly "errors" except to a niggardly imagination'. Morgan's essay, like Austin's felicitous account of the 'infelicities', is a demonstration in the comic mode: 'It may seem extraordinary that such brilliant compressions of relevant meaning . . . should all start up unbidden out of mere confusion and hastiness of mind.' It is also, despite the philosophical irony motivating its patient accumulation of data about actual usage, basically optimistic in its view of the workings of spontaneity, the 'self unqualified in volition', and the complex and recalcitrant nature of things. 'Brilliant' is of course the travesty of a felicity, a cleverly planted self-stultifying oxymoron. A brilliant self-betrayal must be in a peculiar way hollow or void, a kind of joking in which the agent fails to see the joke through which our 'murky thought willy-nilly comes clean', is 'betrayed into

157

actuality'. The 'first-grade or classic pun' is parenthetically insulated by the imbecile context and it gives 'pleasure to the reader' somewhat in the way that Sir Fopling Flutter and Mrs Malaprop gave pleasure to the polite audiences of their times. It is not unlike our relish of the fine distinction between 'clumsily he trod on the snail' and 'he trod on the snail clumsily', where it would be disproportionate to consider the feelings of the snail. Actuality is something brute; but romantic pessimism and melodramatic addiction to culpability are still subject to Iris Murdoch's caveat: 'ideas of guilt and punishment can be the most subtle tool of the ingenious self . . . and the unworthiness of one's motives is interesting'. Yet here again distinctions may validly be drawn. 'Motives', in Miss Murdoch's critical usage, are transcendental, are part of the solipsistic 'interestingness of existence' and do not necessarily equate with vocational, technical intent. Greville's 'creeping Genius' is something other than the melancholy poor cousin of Murdoch's 'ingenious self' and Sidney's distinction between 'erected wit' and 'infected wil' is not overly concerned with 'ideas of guilt and punishment'. The essential question is whether one can properly talk of an 'irreproachable skill'. Pound thought that one could. 'Verbalism demands a set form used with irreproachable skill.' And Kenneth Burke's 'workmanship . . . in which the ethical and the esthetic are one' presupposes that the 'faultless' is in all senses practicable. The ethical and the aesthetic come together at those points where 'freedom of pitch' and 'freedom of field' perfectly intersect or perfectly coincide. And when the conjunction is bungled we discover the complicity between a solecism and 'a sloppy and slobbering world'. The charge that the 'transition from generalization to practice completely eluded' Pound, though excessive, at least takes with a proper seriousness the idea that 'rhetoric' is a part of the ontology of moral action. The desperation of 'I never did believe in Fascism, God damn it', the angry bewilderment of 'everyone's inexactitude very fatiguing', are both pre-judged by 'the tyro can not play about with such things, the game is too dangerous'. Pound had written this, in 1917, in an essay on Laforgue, 'the finest wrought' of modern French satirists. 'Finest wrought' and 'everyone's inexactitude' are mutually uncomprehending and Pound stands condemned by his own best judgment, the 'tyro' to his own mastery. The transcript of the Washington hearing preserves a number of solemn and vacuous pronouncements by advocates and experts on both sides, but the observation that 'the crime with which he is charged is closely tied up with his profession of writing' has an ineluctability that is not diminished by its banal obviousness. Austin's principles are vindicated, though his prejudices and self-satisfactions are not. The word-monger, word-wielder, is brought to judgment *'by his being the person who does the uttering . . . In written utterances (or "inscriptions"), by his appending his signature'*. Our

158

word is our bond. It was suggested, at the start of this discussion, that Pound may have misconstrued a fine point of semantics. In *How To Do Things With Words* Austin writes that 'a verdictive is a judicial act as distinct from legislative or executive acts, which are both exercitives'. Pound's error was to confuse the two, to fancy that poets' 'judicial sentences' are, in mysterious actuality, legislative or executive acts. But poets are not legislators, unless they happen to be so employed, in government or law; and '*recta ratio factibilium*' is not '*mens rea*' or '*actus reus*'. The 'world's' revenge, during his court-hearing and its aftermath, was unwittingly to pay him back, confusion for confusion, with legislative or executive acts presuming to be true verdictives: 'Now comes Dorothy Pound, as Committee of the person and estate of Ezra Pound, an incompetent person . . .' If we seek the mot juste, it was no more and no less than poetic justice. 'And when one has the mot juste, one is finished with the subject.'

Reference Notes

The numerals at the left refer to page numbers in this book. With certain exceptions which I trust are self-evident, the words that follow the numerals indicate the beginning and end of the annotated phrase or passage.

1. *Poetry as 'Menace' and 'Atonement'*

1 *Thus my noblest . . . menace.*/Karl Barth, *The Epistle to the Romans*, Translated from the sixth edition by Edwyn C. Hoskyns (London, 1933), 266.

  'lesse suttle . . . passionate'/*Complete Prose Works of John Milton*, 7 vols (New Haven, 1953–80), II, 403.

2 'mighty figures . . . marble'/Michael Meyer, *Ibsen*, paperback edition (Harmondsworth, 1974), 244.

  'a poem . . . box'/*Letters on Poetry from W. B. Yeats to Dorothy Wellesley* (London, 1940), 24.

  when the words . . . indescribable./T. S. Eliot, *On Poetry and Poets* (London, 1957), 98.

3 'The poet's job . . . justice'/*The Letters of Ezra Pound 1907–1941*, ed. D. D. Paige (London, 1951), 366.

  'the effort . . . verse'/W. Empson, *Seven Types of Ambiguity*, third edition, revised (London, 1953), ix.

  'method . . . posed'/E. Auerbach, *Mimesis*, new edition, paperback (New York, 1957), 356.

  'a concretion . . . ultimately are'/D. M. MacKinnon, *The Problem of Metaphysics* (London, 1974), 110.

  Those masterful . . . heart./*The Collected Poems of W. B. Yeats*, second edition, 1950 (London, 1952), 391.

  'poem . . . sayable'/Henry Rago, 'Faith and the Literary Imagination – The Vocation of Poetry', in Nathan A. Scott Jnr (ed.), *Adversity and Grace: Studies in Recent American Literature* (Chicago and London, 1969), 251.

4 'Poetry . . . real ones'/*The Notebooks of Samuel Taylor Coleridge*, ed. Kathleen Coburn, 3 vols (New York, 1957–), I (Text), 90 (87G.81).

  'one . . . philosopher'/Hannah Arendt, *Men in Dark Times*, paperback edition (Harmondsworth, 1973), 153–4.

  'requires . . . success'/Frank Kermode, *Modern Essays* (London, 1971), 95: referring to the theories of Ehrenzweig.

  'Surrealism . . . against them'/Jean-Paul Sartre, 'Black Orpheus', translated by Arthur Gillette, *Stand*, 6, no. 1, p. 9.

161

'Her . . . these'/Hugh Kenner, *The Pound Era,* paperback edition (London, 1975), 175.

'confession . . . through'/Thomas Mann, *The Genesis of a Novel,* translated by Richard and Clare Winston (London, 1961), 73: 'Bekenntnis und Lebensopfer durch und durch'.

'"angelism . . . order'/Nathan A. Scott Jnr, *The Broken Center: Studies in the Theological Horizon of Modern Literature* (New Haven, 1966), 53–4.

'does not sell . . . holiness'/J. Grotowski, *Towards a Poor Theatre,* ed. Eugenio Barba, 1968, paperback edition (London, 1975), 34.

5 'the mind's . . . thinking'/*Biographia Literaria* by S. T. Coleridge, ed. J. Shawcross, 2 vols (Oxford, 1907), I, 86.

'win[ning] . . . stream'/ibid.

'human . . . intellect'/ibid., 93.

'return . . . upon himself'/*The Complete Prose Works of Matthew Arnold,* ed. R. H. Super, 11 vols (Ann Arbor, 1960–77), III, 267.

'There was . . . himself'/Stuart M. Sperry, *Keats the Poet* (Princeton, NJ, 1973), 252.

'tortuous . . . reader'/MacKinnon, *Problem of Metaphysics,* 66.

Not . . . be./*The Poems of Gerard Manley Hopkins,* ed. W. H. Gardner and N. H. MacKenzie, paperback issue of the fourth (corrected) edition (London, 1970), 99.

6 'the coinherence . . . culture'/Nathan A. Scott Jnr (ed.), *The New Orpheus: Essays toward a Christian Poetic* (New York, 1964), 161

'the principal . . . scientism'/ibid., 142.

no more than a structure of grammar and syntax/ibid., 160.

'intransitive attention'/Nathan A. Scott Jnr, *Negative Capability: Studies in the New Literature and the Religious Situation* (New Haven, 1969), 97.

'a calculated . . . meditation'/*The New Orpheus,* 147; cf. p. 63.

'the world . . . work'/ibid., 163.

'makes good . . . experience . . .'/ibid.

'adventure in atonement'/Scott (ed.), *Adversity and Grace,* 50.

'the comedy . . . redemption'/ibid., 55.

'the drama . . . reconciliation'/ibid., 38.

'the language . . . things'/D. M. MacKinnon, *A Study in Ethical Theory* (London, 1957), 138.

'when the language . . . reduction'/Rago, in Scott (ed.), *Adversity and Grace,* 242

'the empirical guilty conscience'/John H. Rodgers, *The Theology of P. T. Forsyth* (London, 1965), 40.

7 'a saint . . . the soup'/G. K. Chesterton, *Charles Dickens* (London, 1906), 36.

For one . . . mistake./Helen Waddell, *Peter Abelard,* popular edition (London, 1939), 23–4.

Dr Matthew Hodgart/'Misquotation as Re-creation', *Essays in Criticism,* III (1953), pp. 28ff.

Professor Christopher Ricks/'Pater, Arnold and Misquotation', *TLS,* 25 Nov. 1977, pp. 1383–5.

'No man . . . money.'/*Boswell's Life of Johnson,* ed., G. B. Hill, revised and enlarged edition by L. F. Powell, 6 vols (Oxford, 1934), III, 19.

'numerous . . . literature'/ibid., 19–20.

'A knitting editor . . . the other".'/Katherine Whitehorn, in *The Observer Review*, 6 Oct. 1968, p. 27.

8 'anybody . . . courts'/Simone Weil, *The Need for Roots*, translated by A. F. Wills (London, 1952), 36–7.

'social and public institution'/Donald Davie, in *PN Review*, 5, no. 1, 1977, p. 18.

'the fullness . . . medium'/ W. K. Wimsatt, *The Verbal Icon*, first UK edition (London, 1970), 269.

Stephen Spender . . . thought./Spender, *Eliot* (London, 1975), 188.

It's harder . . . sinner./T. S. Eliot, *The Elder Statesman: A Play* (London, 1959), 89–90.

'the point . . . "crime".'/Spender, op. cit., 188.

It may be . . . anxiety./H. Guntrip, *Psychology for Ministers and Social Workers* (London, 1949), 84.

'pathological . . . impotence'/Delmore Schwartz, 'T. S. Eliot as the International Hero', in *Selected Essays of Delmore Schwartz*, ed. D. A. Dike and D. H. Zucker (Chicago and London, 1970), 124–5.

9 'Claverton . . . gesture'/Grover Smith, *T. S. Eliot's Poetry and Plays*, second edition (London, 1974), 248.

'the kind . . . gives'/Eliot, *On Poetry and Poets*, 18.

the ascetic . . . utility./ibid., 85.

'abstention'/ W. W. Robson, *Critical Essays* (London, 1966), 49.

'a fringe . . . express'/Eliot, *On Poetry and Poets*, 86–7.

'as he grew . . . truthtelling'/*New Statesman*, 17 Sept. 1976, p. 376.

10 'the local . . . shape'/ibid.

'struggle . . . "poetry-lovers"'/C. K. Stead, *The New Poetic*, paperback edition (Harmondsworth, 1967), 189.

'it is not . . . deafness'/Jon Silkin, in *The Review*, ed. Ian Hamilton, nos 29–30, p. 10.

'Where . . . poets?'/Keidrych Rhys, *Poems from the Forces* (London, 1941), xiii: 'Where are our war poets? . . . That rhetorical question one has so often heard asked by our Sunday newspapers and public men!'

[Rupert Brooke] . . . attitude./*The Letters of Charles Hamilton Sorley* (Cambridge, 1919), 263. John H. Johnson, *English Poetry of the First World War* (Princeton and London, 1964), 35, and Jon Silkin, *Out of Battle: the Poetry of the Great War* (London, 1972), 75, quote and comment upon this letter.

11 A voice . . . any other./Quoted in *The Collected Poems of Rupert Brooke*, second edition (London, 1928), clvii.

'shot . . . Hulluch'/Sorley, *Letters*, 12.

For . . . something./Rush Rhees, *Without Answers* (London, 1969), 150.

12 it is precisely . . . success./Stephen Prickett, 'True Education must create an accountable élite', *THES*, 17 June 1977, p. 15.

In silence . . . the moon./*Coleridge: Poetical Works*, ed. E. H. Coleridge. New edition (London, 1967), 408.

Humphry House/*Coleridge* (London, 1953), 101.

In his loneliness . . . arrival./Coleridge, *Poetical Works*, 197.

13 The two . . . in doubt./Kathleen Coburn, *The Self Conscious Imagination* (London, 1974), 1.

'prophetic . . . protesting'/Sunday O Anozie, *Christopher Okigbo* (London, 1972), 181.

'the political . . . God'/Cesare Pavese, *This Business of Living*, translated by A. E. Murch, paperback edition (London, 1964), 103: 'Il corpo politico non muore e non risponde quindi davanti a nessun dio'.

'a clear . . . rebel'/Quoted by M. I. Finley, *The Listener*, 5 June 1969, p. 790.

'American . . . morality'/Conor Cruise O'Brien, *The Suspecting Glance: The T. S. Eliot Memorial Lectures delivered at Eliot College in the University of Kent at Canterbury, November 1969* (London, 1972), 34–5.

'the famous . . . appropriate'/ibid., 35.

'understood . . . change'/ibid., 47.

14 'conceptual . . . judgment'/Adrian Cunningham, 'Salvation through Art', in Ian Gregor and Walter Stein (ed.), *The Prose for God:Religious and Anti-Religious Aspects of Imaginative Literature* (London, 1973), 24.

'suggestion . . . actual'/MacKinnon, *A Study in Ethical Theory*, 188.

'the study . . . science'/O'Brien, op. cit., 90.

'slogans . . . sages'/ibid., 11.

'technically sweet'/Robert Oppenheimer, quoted by Robert Jungk, *Brighter Than a Thousand Suns*, paperback edition (Harmondsworth, 1960), 266.

'one can never . . . utilitarianism'/MacKinnon, *A Study in Ethical Theory*, 5–6.

15 'it is both . . . property'/W. H. Auden, *The Dyer's Hand and Other Essays* (London, 1963), 23.

'since there have bene . . . Poets'/*The Prose Works of Sir Philip Sidney*, ed. Albert Feuillerat, 4 vols (Cambridge 1912, reprinted 1962), III, 10.

'the poet . . . lighter ones'/*Conversations with Claude Lévi-Strauss*, ed. G. Charbonnier, translated by John and Doreen Weightman (London, 1969), 111.

'specific gravity . . . as such'/Quoted by C. J. Shebbeare, 'The Atonement and Some Tendencies of Modern Thought', in L. W. Grensted (ed.), *The Atonement in History and in Life* (London, 1929), 302: 'Die Sünde ist das spezifische Gewicht der menschlichen Natur als solcher'.

'searching . . . cosmos'/Keith Sagar, *The Art of Ted Hughes* (London, 1975), 4.

'high energy-construct . . . discharge'/Charles Olson, 'Projective Verse', in James Scully (ed.), *Modern Poets on Modern Poetry* (London, 1966), 272.

'Baudelaire . . . damnation'/T. S. Eliot, *Selected Essays*, second edition revised and enlarged (London, 1934), 391.

'frightful discovery of morality'/ibid., 163.

16 In every age . . . damned./ibid.

'real poets . . . verse'/Alexander Gladkov, *Meetings with Pasternak*, translated and edited by Max Hayward (London, 1977), 155.

'somewhat of the opinion . . . work'/Charles Péguy, *Basic Verities*, rendered into English by Ann and Julian Green (London, 1943), 24.

'the academic . . . pray'/H. A. Williams, 'Theology and Self-Awareness' in A. R. Vidler (ed.), *Soundings* (Cambridge, 1963), 71.

'Attrition . . . no more'/Quoted by Philip Martin, *Mastery and Mercy* (London, 1957), 95. cf. *The Sermons of John Donne*, ed. E. M. Simpson and G. Potter, 10 vols (Berkeley and Los Angeles, 1953–1962) IX, 266: 'For, for Contrition, we doe not, we dare not say, as some of them, that Attrition is sufficient . . .'

'repentance . . . *(meta-noia)*'/Fr Martin Jarrett-Kerr, CR, *Our Trespasses: A Study in Christian Penitence* (London, 1948), 97.
After . . . redemption./Wallace Stevens, *Opus Posthumous* (London, 1959), 158.
'explicitly . . . poetry'/Vincent Buckley, *Poetry and Morality* (London, 1959), 41.
17 'laboriously acquired' . . . clearness'/Arthur Symons, *The Symbolist Movement in Literature*, revised edition 1919 (New York, 1958), 82, 66, 40.
'must . . . language'/Scott, *The Broken Center*, 67.
'the effect . . . despair'./P. T. Forsyth, *The Principle of Authority*, (London, n.d.), 7–8.
'The man . . . popular'/ibid., 324.
'being fallen into the "they"'/Dietrich Bonhoeffer, *No Rusty Swords*, translation revised and edited by John Bowden in conjunction with Pastor Bethge, paperback edition (London, 1970), 51: quoting Heidegger.
18 'word-helotry'/George Steiner, 'In Bluebeard's Castle', *The Listener*, 15 April 1971, p. 476.
To take . . . to praise./From William Empson, 'Courage Means Running', *Collected Poems* (London, 1955), 56–7.

2. *The Absolute Reasonableness of Robert Southwell*

19 'inhuman ferocity'/*Publications of the C[atholic] R[ecord] S[ociety]* (London, 1905–), vol. v: J. H. Pollen, SJ (ed.), *Unpublished Documents Relating to the English Martyrs*, 1 (1908), 325. Pollen's translation of Southwell's Latin.
'Grinding . . . punishments'/Robert Southwell, *An Humble Supplication to Her Maiestie*, ed. R. C. Bald (Cambridge, 1953), 34
'more fierce and cruell'/*CRS* v, 208: Richard Verstegan's Dispatches.
'to the reverend . . . Fayth.'/*An Epistle of Comfort, To The Reverend Priestes, & to the Honorable, Worshipful, & other of the Laye sort restrayned in Durance for the Catholicke Fayth . . . Imprinted at Paris*. See Pierre Janelle, *Robert Southwell the Writer*,' reprint of the 1935 UK edition (Mamaroneck, NY, 1971), 310: 'really printed, perhaps, at a secret press in England' (London? 1587–8).
'seeing . . . contemplate'/J. R. Roberts, 'The Influence of *The Spiritual Exercises* of St Ignatius on the Nativity Poems of Robert Southwell', *Journal of English and Germanic Philology*, LIX (1960), p. 452; cf. Janelle, op. cit., 109.
'almost inevitable martyrdom'/*Humble Supplication*, xvii. Editorial introduction.
'yearning . . . purpose'/A. O. Meyer, *England and the Catholic Church under Queen Elizabeth*, authorized translation by J. R. McKee (London, 1916), 190, 212.
'there is a note . . . methodology'/Roberts, op. cit., p. 455; cf. Janelle, op. cit., 110, 'mystical outpourings'; Louis L. Martz, *The Poetry of Meditation*, revised edition (New Haven and London, 1962), 36, 69, 78, 83.
20 'bless . . . objects'/*CRS* v, 319: Pollen's translation of Southwell's Latin.
'particularly . . . Preachers'/[Richard Challoner], *Memoirs of Missionary Priests* ([London] 1741), 42. Brotherton Library, University of Leeds: Special Collections.
'Every priest . . . matter'/*CRS* v, 318, 316: Pollen's translation of Southwell's Latin.
'English . . . tongue'/*The Poems of Robert Southwell, SJ*, ed. James H. McDonald and Nancy Pollard Brown (Oxford, 1967), xix–xx.

'an England . . . Puritan'/Martz, op. cit., 9.

'the internal . . . camp'/Meyer, op. cit., 171.

'mutual recrimination' . . . laity';/John Bossy, *The English Catholic Community 1570–1850* (London, 1975), 32.

'amongst . . . kind'/J. C. H. Aveling, *The Handle and the Axe: the Catholic Recusants in England from Reformation to Emancipation* (London, 1976), 68.

'special . . . legend'/ibid., 72, 67.

'a cheerful . . . manliness'/Janelle, op. cit., 3.

'mild . . . temper'/Martz, op. cit., 205.

'in great favour . . . respect'/Janelle, op. cit., 109–10.

21 'Campion's Brag'/cf. A. C. Southern, *Elizabethan Recusant Prose 1559–1582* (London, 1950), 151, 153–6.

'rapidly . . . journey'/*Humble Supplication*, xvii; Richard Simpson, *Edmund Campion: A Biography*, new edition 'reprinted from a copy corrected by the learned Author before his death' (London, 1896), 225.

'spontaneous effusions'/Janelle, op. cit., 109.

'well-ordered will'/ibid., 111. The phrase is apt; though more recent scholarship seriously doubts Southwell's authorship of the translation from Estella.

'the deliberate . . . poetry'/*The Sermons and Devotional Writings of Gerard Manley Hopkins* ed. Christopher Devlin, SJ (London, 1959), 118.

'suitable . . . expression'/Janelle, op. cit., 153.

accusations and slanders/See Simpson, op. cit., 224–5.

The 1591 . . . Ruffians . . .'/*Humble Supplication*, xi, 60, 64.

'*miles gloriosus*'/Simpson, op. cit., 365, 367.

'the most . . . controversy'/Meyer, op. cit., 210.

'the creative art of denigration'/Helen C. White, 'Some Continuing Traditions in English Devotional Literature', in *Publications of the Modern Language Association*, 57, ii (1942), p. 966.

'a spirite . . . devotion'/Quoted in Martz, op. cit., 8; cf. White, op. cit., p. 967.

'uncompromising . . . writings'/R. C. Bald, 'Donne and Southwell', in *Humble Supplication*, 79.

The title-page . . . thoroughly contentious./*The Nevv Testament of Iesus Christ, translated faithfvlly into English* . . . Printed at Rhemes . . . 1582. Brotherton Library, University of Leeds: Special Collections. For example, p. 182: '[Heretikes] may by penal lawes be co[m]pelled to the Catholike faith.'

22 'Controversy . . . poets too'/Meyer, op. cit., 221.

'Witty . . . feeling'/*CRS* v, 90–1.

'a terrible . . . catholics'/Meyer, op. cit., 222.

'caddish'/C. C. Martindale, S J, 'Edmund Campion', in Maisie Ward (ed.), *The English Way* (London, 1933), 240: 'In Elizabeth's time the State became the cad as such'.

'the intimate . . . fields'/David Mathew, 'John Fisher', ibid., 208.

'hankered'/Bossy, op. cit., 11.

'local . . . instances of this./Christopher Haigh, *Reformation and Resistance in Tudor Lancashire* (Cambridge, 1975), 54, 64, 85, 145.

'sluggishness . . . ignorantia'/Bossy, op. cit., 102, 223; cf. *CRS* xxxix (1942), *Letters and Memorials of Fr. Robert Persons*, I, 108.

Elizabethan . . . sank into them,/See Aveling, op. cit., 74, 151–3.

'homo sordidissimus'/Henry Garnet, SJ, quoted in C. Devlin, *The Life of Robert Southwell Poet and Martyr* (London, 1956), 210.

'a man . . . mind'/*John Gerard: The Autobiography of an Elizabethan*, translated from the Latin by Philip Caraman (London, 1951), 230.

'the most desperate . . . equably remarked./H. Thurston, 'Father Southwell the Euphuist', in *The Month*, 83 (1895), p. 243.

'a wayward . . . scapegrace'/Janelle, op. cit., 55; cf. Devlin, *Life of Southwell*, 11, 74, 257n.

23 'ardent . . . appreciations'/Thurston, op. cit., p. 243.

'association . . . England./Simpson, op. cit., 223.

'that wicked . . . mischief'/*CRS* v, 314. Pollen's translation of Southwell's Latin.

'greene witts . . . witt'/*Humble Supplication*, 18.

'apparell . . . Calling'/ibid., 8.

'garnished . . . colours'/Caraman, op. cit., 18n.

Anthony Copley . . . between his teeth/Devlin, *Life of Southwell*, 257n.

os impudens/Simpson, op. cit., 368.

Dr William Allen . . . vernacular./Janelle, op. cit., 7–9 and 9 n. 20. T. F. Knox's translation of Latin MS, English College, Rome.

'Southwell . . . Puritans'/ibid., 198.

24 'a thoroughly . . . religion'/Ernst Cassirer, *The Platonic Renaissance in England*, translated by J. P. Pettegrove (London, 1953), 74.

Christopher Morris . . . Elizabeth herself'/Christopher Morris, *Political Thought in England: Tyndale to Hooker* (London, 1953), 174–5, 126, 195.

'no puritan . . . hatred'/Meyer, op. cit., 350.

'contumelious termes'/*Humble Supplication*, 2.

'a reply . . . scorn'/Meyer, op. cit., 351.

Edmund Campion . . . equitie'./Southern, op. cit., 154–5.

The Proclamation . . . pecuniary summe . . .'/*Humble Supplication*, 60–1, 63.

'measure . . . Equity'/ibid., 3.

'Sound beliefe . . . disposition./ibid., 14.

25 'Let it be scanned . . . carrieth'/ibid.

'For a long time . . . should be'/*Selected Historical Essays of F. W. Maitland*, chosen and introduced by Helen M.Cam (Cambridge, 1957), 133, 131. 'Its' refers to the Court of Chancery which dispensed equity.

'notion . . . law"'/L. L. Fuller, *Legal Fictions* (Stanford, California, 1967), 87.

In the beautiful . . . himself'./*Marie Magdalens Funeral Teares* [1591] by Robert Southwell, a Facsimile Reproduction with an Introduction by Vincent B. Leitch (Delmar, NY, 1975), 32 and 32 verso (henceforward *MMFT*).

If equities . . . weightes:/*Poems*, ed. cit., 75.

'Thogh I were . . . disprove my courses'./*MMFT*, 39 verso; 40 verso; *Humble Supplication*, 2-3; Robert Southwell, SJ, *Two Letters and Short Rules of a Good Life*, ed. Nancy Pollard Brown (Charlottesville, Va., 1973), 33, 79.

26 the 'great age . . . dilemma'/H. R. Trevor-Roper, *Historical Essays* (London, 1957), 92.

'uncanonical'/Meyer, op. cit., 79.

'declaration . . . priest'/Bossy, op. cit., 37.

And for the next . . . scaffold/*CRS* v, 8–17.

as Christopher Morris has argued/op. cit., 106.

'Because . . . reason'/Devlin, *Life of Southwell*, 287.

Maitland has said/Maitland, op. cit., 125.

'during Shakespeare's . . . equity'/J. W. Dickinson, 'Renaissance Equity and *Measure for Measure*', in *Shakespeare Quarterly*, XIII (1962), p. 287; W. Dunkel, 'Law and Equity in *Measure for Measure*', ibid., p. 285.

27 As Paul Vinogradoff has shown . . ./*Roman Law in Medieval Europe*, second edition (Oxford, 1929), 55–6, 66; see also 106: 'justitia est constans et perpetua voluntas jus suum cuique tribuendi. [the definition of justice as given in the Digest]'.

'a work . . . beauty'/Janelle, op. cit., 286–7.

'great plea . . . silence'/R. W. Chambers, *Thomas More*, second edition (London, 1938), 336.

'monster . . . taciturnity'./*CRS* v, 283.

'a new kind . . . one word'/Janelle, op.cit., 66–7.

'hinterland'/This term was suggested to me by D. W. Harding's brilliant essay, 'The Hinterland of Thought', in *Experience into Words*, paperback edition (London, 1974), 176ff. In fairness to Harding and myself it must be added that I employ it in a different sense.

'wittye crueltye'/*Epistle of Comfort*, 125 verso, translating St Cyprian.

The executioner . . . porredge pot./*The Works of Thomas Nashe*, Edited from the original texts by Ronald B. McKerrow, reprinted from the original edition with corrections and supplementary notes edited by F. P. Wilson, 4 vols (Oxford, 1958), II, 327.

28 'the pleasure . . . description'/G. R. Hibbard, *Thomas Nashe: A Critical Introduction* (London, 1962), 174.

'the martyr . . . his mouth"'/*CRS* v, 207.

And as a cuninge . . . by them./ed. cit., 203 recto, verso.

'Our teares . . . felicitye'/ibid., 113 recto.

It is a great . . . our cure./Bd. Robert Southwell, SJ, *Spiritual Exercises and Devotions*, edited by J.-M. de Buck, SJ, and translated by P. E. Hallett, abridged edition, English text only (London, 1974), 29. Hallett's translation of Southwell's Latin.

'had particular significance for Southwell'/*Poems*, ed. cit., 146.

29 The modern 'consensus' . . . Christ./F. V. Filson, *A Commentary on the Gospel According to St. Matthew* (London, 1960), 138; D. Hill (ed.), *The Gospel of Matthew* (London, 1972), 200; J. C. Fenton, *The Gospel of St. Matthew* (Harmondsworth, 1963), 179–80. Hopkins reads the text in Southwell's spirit. See *The Sermons and Devotional Writings of Gerard Manley Hopkins*, ed. Christopher Devlin, SJ (London 1959), 96.

epigraph on the title-page/ed. cit. cf: 'And from the dayes of Iohn the Baptist vntil novv, the kingdom of heauen suffereth violence, and the violent beare it avvay.' *Rhemes NT, 1582*, ed. cit., 29.

'And though . . . foughte for it'/ed. cit., 31 verso, 32 recto.

Thy ghostly . . . tender love./*Poems*, ed. cit., 55.

'in my judgment . . . to God'/Luther, *Lectures on Romans*, newly translated and edited by W. Pauck, The Library of Christian Classics xv (London, 1961), 349.

'if you wish . . . fear nothing'/*Spiritual and Anabaptist Writers*, ed. G. H. Williams and A. M. Mergal, The Library of Christian Classics xxv (London, 1957), 369.

'Thus . . . knives'/*The Great Commentary of Cornelius à Lapide*, translated by T. W. Mossman (London, 1876–97): *St Matthew's Gospel, chaps X to XXI*, 58, 60. I am grateful to Fr Martin Jarrett-Kerr CR for discussing this and other questions with me though I am, of course, solely responsible for any errors of fact or interpretation.

30 'tiranical persecution . . . thy desire'/*Epistle of Comfort*, ed. cit., 79 verso; *MMFT*, ed. cit., 25 recto; *Humble Supplication*, ed. cit., 4; *MMFT*, 58 recto; ibid., 7 recto; ibid., 38 recto, verso.
'I am urged . . . detest'/*Two Letters . . .* , ed. cit., 56.
[A]s not . . . spiritual plane'./Southwell, *Triumphs over Death*, quoted Janelle, op. cit., 233. Janelle's comment, 232.
'Of the venerable martyr . . . butchery'./*The Rambler*, NS VIII (1857), 114, quoted in Dom Bede Camm, *Forgotten Shrines*, second edition (London, 1936), 359 and 361n.

31 The 'officers . . . George Haydock./*CRS* v, 288–9, 57–62.
For Southwell . . . emblem/cf. J. Morris, SJ, *The Troubles of our Catholic Forefathers related by themselves*, First Series (London, 1872), 98: (the execution of Fr Cuthbert Maine, 1587) 'a very high gibbet . . . and all things else, both fire and knives, set to the show and ready prepared'.
In a letter . . . medicines"'./*CRS* v, 318. Pollen's translation of Southwell's Latin; p. 301, Pollen's note on 'veiled style'.
Then sayd the Angel . . . them./*The Holie Bible Faithfvlly Translated into English . . .* (Douai, 1609–10), 998.

32 Remember . . . dyed./*Epistle of Comfort*, ed. cit., 201 recto.
indeed . . . martyrdom./ibid., 202 verso. For a caveat on 'Catholic major gentry portraits' see Aveling, op. cit., 150–4.
'the reparation . . . offences'/Maitland, op. cit., 131.
'transformation'/cf. Janelle, op. cit., 190; Helen C. White, 'Southwell: Metaphysical and Baroque', in *Modern Philology*, 61(1963–4), p. 161; cf. F. W. Beare, *A Commentary on the Epistle to the Philippians*, third edition (London, 1973), 140: 'Even now [Paul] tells us, the transformation is proceeding within us'.
'translations . . . alteratio[n]'/Janelle, op. cit., 184; *Poems*, ed. cit., xcvi, 135; *Epistle of Comfort*, 154 recto.
And this . . . glorye./*Epistle of Comfort*, 203 verso, 204 recto.
[He] . . . glory'/Beare, *Commentary*, 138.
Christ's Parousia/Cf. Beare, 138, 140.

33 There is no reason . . . the Crosse./*Epistle of Comfort*, 32 recto, verso.
'transfigured . . . passion'./ibid., 32 recto.
'Calvary's turbulence'/'The Magi': *The Collected Poems of W. B. Yeats*, second edition, with later poems added (London, 1950), 141.
'For Thy sake . . . nothing'./*Spiritual Exercises and Devotions*, 74: Hallett's translation of Southwell's Latin.
a fair amount of Ovid by heart/Janelle, op. cit., 135.
'lingering . . . skill'/I apply a phrase of Hopkins, 'Deutschland', st. 10./*Poems*, ed. Gardner and MacKenzie (London, 1970), 54.
'From bosomes . . . tumble thick . . ./*Seneca His Tenne Tragedies translated into Englysh* (London, 1581), 33 recto (*Thyestes*), 72 recto (*Hippolytus*).

34 Whye doest . . . reason./*Epistle of Comfort*, 205 recto.

'masculine perswasive force'/'On his Mistris', John Donne, *The Elegies and the Songs and Sonnets*, ed. Helen Gardner (Oxford, 1965), 23.

'make . . . experience'/ibid., xviii

'how well . . . together'/'The Author to his loving Cosen', *Poems*, ed. cit., 1.

'Whose measure . . . too little have'/*Poems*, ed. cit., 2, 5, 28.

'Baroque . . . transformation'/White, loc. cit.

'uneven accompt' . . . 'just measure':/*Poems*, ed. cit., 79, 77.

'the affections . . . God'./*Epistle of Comfort*, 191 recto, 190 verso.

35 'loathed pleasures . . . Cruell Comforts'/*Poems*, ed. cit., 50, 41, 54; *Humble Supplication*, 34.

'awestruck . . . man'/Roberts, op. cit., 454.

This little Babe . . . surprise./'New heaven, new warre', *Poems*, ed. cit., 14.

'naïve . . . concession'/*Agenda*, 13, no. 3 (Autumn 1975), 32. For Southwell's phrase see *MMFT*, ed. cit., p. A3 verso.

'complexity of association'/White, op. cit., p. 166.

36 Antonin Artaud . . . Rodez/Charles Marowitz, *Artaud at Rodez* (London, 1977), 73.

'Excessus' signifies 'ecstasy'./Etienne Gilson, *The Mystical Theology of Saint Bernard*, translated by A. H. C. Downs (London, 1940), 26. See also Gordon Rupp, *The Righteousness of God* (London, 1953), 143: "'Excessus mentis," [Luther] says, means "either the alienation of mind . . ." or "the rapture of the mind into . . . faith," and this is really what is meant by "ecstasis".' *A Study of Wisdom: Three Tracts by the Author of 'The Cloud of Unknowing'*, translated into Modern English . . . by C. Wolters (Oxford, 1980), 21: '"Ibi Benjamin adolescentus in mentis excessu" [Psalm 68:28.Vulgate] which means in English, "There is Benjamin, the young child, transported out of mind".'

'solace' . . . Tower/Janelle, op. cit., 68.

'pretty Babe . . . breast'/*Poems*, ed. cit., 15.

'the most hackneyed . . . tradition'/Janelle, op. cit., 168; *Poems*, ed. cit., 124.

'variation . . . manner'/Martz, op. cit., 81–3, 364. Puente's work was first published in 1605. John Heigham's English translation was not issued from St Omer until 1619.

St Ignatius . . . tradition./Janelle, op. cit., 109.

'excess . . . necessity'/*Two Letters . . .* ed. cit., 36.

37 'fancie' . . . 'selfe delight'/*Poems*, ed. cit., 50.

'Let vs . . . therin . . .'/ed. cit., 37 verso.

'the Jesuit discipline . . . wildness . . .' Devlin, *Life of Southwell*, 85.

I would further suggest . . . not lost on him./cf. Etienne Gilson, *The Spirit of Medieval Philosophy*, translated by A. H. C. Downs (London, 1936), 290–2; Rupp, op. cit., 143; C. Wolters, *Three Tracts . . .*, 21.

'I am come hither . . . poor life'./Devlin, *Life of Southwell*, 321.

'Loue . . . loue'./*MMFT*, 52 recto, verso.

3. *The World's Proportion*

38 *The worlds . . . awrie.*/John Donne, 'The First Anniversary', ll.302–4. *The Epithalamions Anniversaries and Epicedes*, ed. W. Milgate (Oxford, 1978), 30.

Jonson . . . Romans . . ./J. E. Sandys in *Shakespeare's England*, 2 vols (Oxford, 1916), I, 274.

'Words are the Peoples'/B[en] J[onson], ed. C. H. Herford and Percy Simpson, 11 vols (Oxford, 1925–52) VIII, 621.

'accurate eye . . . ordinary speech'/Marchette Chute, *Ben Jonson of Westminster* (New York, 1953), 85, 74.

39 the tricks . . . nose./L. C. Knights, 'Ben Jonson, Dramatist' in *The Age of Shakespeare*, ed. B. Ford. Reprinted with revisions (Harmondsworth, 1956), 304.

Character writers . . . stock response./See D. Nichol Smith, *Characters of the Seventeenth Century* (Oxford, 1918), 2, 63, 223. Of James I, 'some *Parallel'd* him to *Tiberius* for *Dissimulation* . . .'; of Charles II, 'His person and temper, his vices as well as his fortunes, resemble the character that we have given us of *Tiberius* . . .'; of Strafford, 'in a worde, the Epitaph . . . that Silla wrote for himselfe, may not be unfitly applyed to him'. See also *Sir Fulke Greville's Life of Sir Philip Sidney etc. First Published 1652* with an introduction by Nowell Smith (Oxford, 1907), 39: 'nor yet like that gallant Libertine *Sylla* . . .'. These polemic analogues share a common field of classical reference with Jonson's two plays.

CICERO . . . infamy?/*Catiline*, IV 316ff. *BJ* v 508.

40 We demand . . . stripping us . . ./A. R. Bayley, *The Great Civil War in Dorset* (Taunton, 1910), 351–3. For the discovery of this passage I am indebted to Christopher Hill and Edmund Dell (ed.), *The Good Old Cause* (London, 1949), 368. However, their transcript differs from Bayley's in certain details. I have retained Bayley's readings.

'shipwrack'd . . . fortunes'/*Catiline*, IV, 413ff. *BJ* v, 511.

CAT. . . . stand for./*Catiline*, I, 409ff. *BJ* v, 448.

'industrie', 'vigilance'/*Catiline*, III, 33. *BJ* v, 469.

41 'to make . . . choice'/*Catiline*, II, 373. *BJ* v, 467.

PETREIVS . . . world./*Catiline*, v, 11ff. *BJ* v, 527.

altar . . . bower. . ./'London, 1802', from 'Poems Dedicated to National Independence and Liberty'.

MY LORD '. . . magistrate./Dedication to *Catiline*, *BJ* v, 431.

42 I thinke meete . . . can be./*Wilson's Arte of Rhetorique 1560*, ed. G. H. Mair (Oxford, 1909), 156.

'great melody'/W. B. Yeats, 'The Seven Sages', *The Collected Poems of W. B. Yeats*, second edition, with later poems added (London, 1950), 272.

SEM. . . . competitors . . ./*Catiline*, II, 96ff. *BJ* v, 457–8.

43 There is the incident . . . pages/*Catiline*, I, 506ff. *BJ* v, 451.

'an outrage to probability'/Quoted in *Coleridge on the Seventeenth Century*, ed. R. F. Brinkley. Reprint (New York, 1968), 645.

CET. . . . dead./*Catiline*, III, 663ff. *BJ* v, 490.

In *Epicoene* . . . natures./See E. B. Partridge, *The Broken Compass* (London, 1958), ch. 7.

the noted . . . world./*Seianus*, I, 216–217, *BJ* IV, 362

44 'needy' . . . 'sloth'/e.g. *Catiline*, I, 161, *BJ* v, 440; III, 715, *BJ* v, 492; IV, 184, *BJ* v, 504; IV, 226, *BJ* v, 505; I, 205, 211, *BJ* v, 441.

All *Rome*. . . . suffer,/*Seianus*, v, 256. *BJ* IV, 446.

CIC. . . . such./*Catiline*, III, 438ff. *BJ* v 483.

45 CAT. . . . *nectar* . . ./*Catiline*, I, 102ff. *BJ* v, 438.

46 SEIANUS . . . lusts./*Seianus*, III, 598ff. *BJ* IV, 412–13.

47 SAB. . . . *Rhodes* . . ./*Seianus*, IV, 163ff. *BJ* IV, 424.

James I's . . . succinctly enough./See G. P. Gooch, *English Democratic Ideas in the Seventeenth Century* (1898). Second edition with supplementary notes and appendices by H. J. Laski (1927). Paperback edition (New York, 1959), 53.

48 'government exists . . . accordingly'/L. I. Bredvold, *The Intellectual Milieu of John Dryden* (Ann Arbor, 1934), 143.

'originally non-dramatic' nature/U. M. Ellis-Fermor, *The Jacobean Drama* (London, 1936), 99–100.

49 The golden laws . . . allowing./H. J. C. Grierson, *The Poems of John Donne*, 2 vols (Oxford, 1912), I, 113–16, prints this poem as 'Elegie XVII'. Helen Gardner in her edition of *The Elegies and The Songs and Sonnets* (Oxford, 1965) places it among the 'Dubia' (pp. 104–6) and questions the attribution of the poem to Donne 'on internal evidence' (p. xlvi). I have slightly rephrased my original 1960 text to take account of this caveat.

'free sword'/*Catiline*, I, 230. *BJ* V, 442.

It could be argued . . . comment./This statement is affected by Professor Gardner's opinion but I have decided to let it stand. Dramatists frequently fell foul of the censors. *Sejanus* itself, despite Jonson's caution, barely scraped past official disapproval. Fulke Greville destroyed a tragedy in manuscript because he thought it 'apt enough to be construed, or strained to a personating of vices in the present Governors, and government' (Greville, op. cit., 156).

50 This *anachronic* . . . amusing./Quoted in Brinkley (ed.), op. cit., 643.

A meere vpstart . . . sweat for't./*Catiline*, II, 119ff. *BJ* V, 458. The whole question of the 'new man' is obviously one of the great thorny problems of the era. It cannot be fully discussed here. The fact remains that Cicero, the *arriviste*, is 'good'; Sejanus, the *arriviste*, 'wicked'; and the necessity to cajole us into the acceptance of this fact influences both action and imagery in the two Roman plays. Cicero is an approved 'new man' but feels some need for self-justification. 'Cecil noted in 1559 that "the wanton bringing up and ignorance of the nobility forces the Prince to advance new men that can serve"' (quoted in *The Age of Shakespeare*, p. 16). The 'new man' versus the 'wanton' nobility is, essentially, the situation of *Catiline*.

the ladie ARETE . . . gowne/*Cynthias Reuells* I, 89ff. *BJ* IV, 38.

we will eate . . . opalls . . ./*The Alchemist*, IV, i. 156ff. *BJ* V, 364.

51 Hence comes . . . made . . ./*Catiline*, I, 573ff. *BJ* V, 453.

CHOR . . . helme . . ./*Catiline*, III, 60ff. *BJ* V, 470.

52 he piercing . . . truth./Greville, op. cit., 36.

We haue . . . for vertue . . ./*Catiline*, II, 131ff. *BJ* V, 458.

He was . . . in others./*Seianus*, I, 124ff. *BJ* IV, 359.

In this . . . Recording Muse./*Absalom and Achitophel*, ll.817–28. *The Poems of John Dryden*, ed. James Kinsley 4 vols (Oxford, 1958), I, 238.

53 NUR . . . crime . . ./*The New Inne*, V, 56ff. *BJ* VI, 486. 'Nurse' is Lady Frampul disguised.

money . . . knaue./*Catiline*, V, 299. *BJ* V, 536.

In *Catiline* . . . tragedy./Editorial 'Introduction to "Catiline his Conspiracy",' *BJ* II, 122.

It is not manly . . . punishment./*The Divell is an Asse*, V, viii, 169ff. *BJ* VI, 269.

4. *The True Conduct of Human Judgment*

55 When Queen Elizabeth . . . masques./E. K. Chambers, in *Shakespeare's England*, 2 vols (Oxford, 1916), I, 105.

especially . . . taste . . ./J. Dover Wilson, prefatory note to *Cymbeline*, ed. J. C. Maxwell (Cambridge, 1960), ix.

'taking rank . . . Yeomen'/F. E. Halliday, *The Life of Shakespeare* (London, 1961), 178.

'since princes . . . things'/Francis Bacon, 'Of Masques and Triumphs'. *The Essays or Counsels, Civil and Moral of Francis Bacon*, ed. S. H. Reynolds (Oxford, 1890), 268.

It has been suggested . . . 'dual purpose' play/J. M. Nosworthy, *Cymbeline*, Arden edition (London, 1955), xvi. This text is used for quotations. A more recent investigator, R. T. Thornberry, in *Shakespeare and the Blackfriars Tradition* (Ohio State University Dissertation, 1964), rejects the dual purpose theory and regards *Cymbeline* as a play composed specifically for the Globe.

'an unlikely . . . tricks'/H. Granville-Barker, *Prefaces to Shakespeare*, Second Series, fourth impression (London, 1944), 244.

'to commit . . . material'/Maxwell, ed. cit., xxxix.

56 the supreme theme . . . destiny/So well described by G. Wilson Knight, *The Crown of Life* (London, 1947), ch. 4.

Emrys Jones's persuasive essay/Emrys Jones, 'Stuart Cymbeline', in *Essays in Criticism*, XI (1961), pp. 84–99.

'the fulfilment . . . people'/ibid., p. 90.

To the medieval chroniclers . . . connotations./R. Moffet, '*Cymbeline* and the Nativity', in *Shakespeare Quarterly*, XIII (1962), pp. 207–18.

I felt that Stalingrad . . . out . . ./Alan Sillitoe, *Road to Volgograd*, paperback edition (London, 1966), 40.

'magnetized . . . spot'/Knight, op. cit., 155.

a kind of interpretative . . . incoherent./Jones, op. cit., p. 98.

'a central fumbling . . . logic'/ibid., p. 97.

'being too close . . . audience'/ibid.

57 Even if it were an established . . . play/It is on record that the play was acted at Court in 1634, but that is not quite the same thing.

'officiously' . . . ill-judged actions'/Jones, op. cit., p. 97.

'two supremely excellent human beings'/Moffet, op. cit., p. 208.

'one of the great women . . . world'/Mark Van Doren, *Shakespeare* (London, 1941), 309.

All I can . . . ought to be . . ./Letter quoted by A. M. Eastman and G. B. Harrison, *Shakespeare's Critics* (Ann Arbor, 1964), 172–3.

('My face . . . appears')/'The Good-morrow', l. 15. John Donne, *The Elegies and The Songs and Sonnets*, ed. Helen Gardner (Oxford, 1965), 70.

58 'quite out of character'/Nosworthy, ed. cit., 17.

'too good to be true'/Maxwell, ed. cit., xxxix.

'Shakespeare intends . . . virtue'/Nosworthy, ed. cit., 4.

That Shakespeare intends nothing . . . embodies./Homer D. Swander, '*Cymbeline* and the "Blameless Hero"', *ELH*, XXXI (1964), pp. 259–70 (p. 260).

59 'the two characters . . . actor'/Homer D. Swander, '*Cymbeline*: Religious Idea and

Dramatic Design', in *Pacific Coast Studies in Shakespeare* (Eugene, Oregon, 1966), 251.

Dr Leavis's objection . . . the play'/*The Common Pursuit* (London, 1952), 176.

Mr Traversi . . . at work here./*Shakespeare: The Last Phase*, second impression (London, 1965), 62.

'the puritanical . . . *The Tempest*'/Knight, op. cit., 149.

60 It is true . . . 'deep centre' here/The words quoted are from Leavis, op. cit., 174.

brancheth itself . . . rumours . . ./Bacon, *The Advancement of Learning*, 1605. *The Works of Francis Bacon* collected and edited by James Spedding . . . 14 vols. New edition (London 1883–92), III, 287–8.

'the sums he gave . . . give'/D. H. Willson, *King James VI and I* (London, 1956), 261.

61 British toughness . . . obnoxious./Knight, op. cit., 136.

The Britons . . . little world'./L. C. Knights, *Drama and Society in the Age of Jonson*, second impression (London, 1951), 245 n. 3. A reference to *Locrine*.

'the character . . . James I'/Jones, op. cit., p. 96.

some earlier editors . . . to 'by'./Moffet, op. cit., p. 215.

62 fearing him as their Iudge . . . teares to God . . ./*The Political Works of James I*, with an introduction by C. H. McIlwain (Cambridge, Mass., 1918), 61.

'the ending of *The Tempest* . . . reticent'/David William, '*The Tempest* on the Stage', in *Jacobean Theatre* (Stratford-upon-Avon Studies 1), ed. J. R. Brown and B. Harris (London, 1960), 135.

'hamartia'/A term covering a gamut of flaws from 'simple error' to 'sin'. For a full description see G. F. Else, *Aristotle's Poetics: The Argument* (Cambridge, Mass., 1957), esp. pp. 376–99. One is aware that Shakespeare may not have known the *Poetics*.

63 If wee present . . . Hodge-podge./I am indebted to the discussion by R. Weimann in *Shakespeare in a Changing World*, ed. A. Kettle (London, 1964), 36–7.

suitable . . . together./Quoted by J. Kerman, *The Elizabethan Madrigal* (New York, 1962), 26.

Their spirit . . . described./Notably by Charles Barber, '*The Winter's Tale* and Jacobean Society', in Kettle, op. cit., 233–52.

it must be confessed . . . life./Bacon, *Works*, III, 396–7.

'the caution . . . judgment',/ibid., 397.

64 'there is more Baconism . . . recognized'/Barber, op. cit., 247–8.

'Experimentation . . . magic'/Hardin Craig, *The Enchanted Glass*, UK edition (Oxford, 1952), 75.

'degenerate Natural Magic'/*Works*, III, 362.

Wilson Knight . . . of her son,/op. cit., 130.

'primal nature . . . aspect'/ibid., 159.

'rehabilitated nature'/cf. Basil Willey, *The Seventeenth Century Background*, sixth impression (London, 1953), ch. 2.

'Man had learned . . . animals'/Craig, op. cit., 99.

65 'on occasion . . . heroine'/Nosworthy, ed. cit., lxii.

'Should not the smoke . . . heavens?'/B. Harris, '*Cymbeline* and *Henry VIII*', in *Later Shakespeare* (Stratford-upon-Avon Studies 8), ed. J. R. Brown and B. Harris (London, 1966), 228.

66 'chromatic tunes'/cf. Peter Warlock, *The English Ayre* (London, 1926), 57.

'Chromatic tunes' is a phrase from a poem set by John Danyel. Warlock analyses the chromaticism of Danyel's setting on pp. 58–61.

'false relation'/False relation is one of the key-themes in *Cymbeline*, implying false conjunction and false report. For a description of false relation in Jacobean music, see the essay by Wilfrid Mellers in *The Age of Shakespeare*, ed. Boris Ford (Harmondsworth, 1956), 394.

Uncertain certain . . . dying last./Quoted in Warlock, op. cit., 57.

'aims at effecting . . . surprise'/H. S. Wilson, '*Philaster* and *Cymbeline*', in *English Institute Essays 1951*, ed. A. S. Downer (New York, 1952), 162.

'the real . . . compromised'/William, op. cit., 135.

Cymbeline . . . 'enchanted ground'./Nosworthy, ed. cit., xlviii.

drenched in flesh . . . conversant . . ./Bacon, *Works*, III, 383.

5. *Jonathan Swift: the Poetry of 'Reaction'*

67 That he was not 'mad' . . . authority/Irvin Ehrenpreis, *The Personality of Jonathan Swift* (London, 1958), 117–26.

68 Swift read . . . Hobbes./H. Williams, *Dean Swift's Library* (Cambridge, 1932), 54 and appendix, 15.

Sir William Temple . . . his pupil./I. Ehrenpreis, *Swift: the Man, his Works, and the Age*, 2 vols (London, 1962, 1967), I, 98.

'more than the "great Ministers" had gone'/W. B. Yeats, *Explorations* (London, 1962), 358.

'occupied so much . . . leisure'/*The Poems of Jonathan Swift*, ed. Harold Williams, 3 vols; second edition (Oxford, 1958), 965.

When I saw . . . your letters./'The Dean to Tho: Sheridan', 1718. *Poems*, ed. cit., 978.

69 I was t'other day . . . twenty years past . . ./*The Correspondence of Jonathan Swift*, ed. Harold Williams, 5 vols (Oxford, 1963–5), V, 271.

'taking his ease at the Castle'/R. Quintana, *Swift: An Introduction* (London, 1955), 11.

Swift has been called . . . wrote'/Yeats, op. cit., 354.

Dr Johnson . . . Dean of St Patrick's./*Lives of the English Poets* by Samuel Johnson. The World's Classics Edition, 2 vols, reset (London, 1952), II, 211–12.

Professor Ehrenpreis . . . his life'./Ehrenpreis, *Swift: the Man* . . . I, 71, 77.

Ireland's ruling caste . . . population'./W. E. H. Lecky, *History of England in the Eighteenth Century*, third edition, revised, 8 vols (London, 1883), II, 221.

70 When to my House . . . Parlor Door . . ./printed in *Poems of Swift*, ed. cit., 1045.

Sheridan's description . . . reciprocating talent./In making this assertion one is obliged to acknowledge, though not to overestimate, Swift's reference to Sheridan's 'bad Verses' (*Correspondence*, II, 307). He had recently been annoyed by the tone of Sheridan's 'The Funeral'.

so successful . . . the other./F. Elrington Ball, *Swift's Verse* (London, 1929), 244.

Swift . . . by heart/Though this may be mythical. For a discussion of Swift in relation to Butler see C. L. Kulisheck, 'Swift's Octosyllabics and the Hudibrastic Tradition', *Journal of English and Germanic Philology*, LIII (1954), pp. 361–8.

Swift's library . . . often associated./Williams, *Swift's Library*, 76: *Paradise Lost*, noted 1715; missing 1745.

'never was one of his Admirers'/*The Prose Works of Jonathan Swift*, ed. Herbert Davis, 16 vols (Oxford, 1939–74), IV, 274.

'agony and indignation'/*Poems by John Wilmot, Earl of Rochester*, ed. V. de Sola Pinto; second edition (London, 1964), xxix.

With mouth . . . those two./Rochester, 'Tunbridge-Wells'. *The Complete Poems of John Wilmot, Earl of Rochester*, ed. David M. Vieth (New Haven and London, 1968), 77.

71 Stand further . . . Matadore./Swift, 'The Journal of a Modern Lady', 1729. *Poems*, 452.

My female Friends . . . *Trumps?*)'/'Verses on the Death of Dr. Swift', 1731. *Poems*, 562.

72 'The Dean . . . the Knave:'/*Poems*, 565.

It is known . . . the description'./Henry Craik, *The Life of Jonathan Swift*, second edition, 2 vols (London, 1894), I, 175; cf. *Poems*, 88–90.

73 Nor do they trust . . . the Man./*Poems*, 450.

'embarrassed . . . situation'/*Poems*, 684: editorial note.

She railly'd well . . . *a Bite.*/*Poems*, 707–8.

74 You must ask . . . a *bite.*/*Correspondence*, I, 40.

The somewhat complex question . . . discussion./See, e.g., A. O. Aldridge, 'Shaftesbury and the Test of Truth', *PMLA*, LX (1945), pp. 129–56; P. Dixon, 'Talking Upon Paper', *English Studies*, XLVI, no. 1 (Feb. 1965), pp. 36–44.

'ironical praise . . . addressed to'/Dixon, op. cit., p. 38. His sources are Steele and Swift. Cf. *Prose Works*, ed. Davis, IV, 91.

So, the pert Dunces . . . Ale./'To Mr. Delany', 1718. *Poems*, 216.

When Swift . . . Raillery'./*Poems*, 890n.

His own poem 'The Journal' . . . friends./*Poems*, 277.

It may seem that infringements . . . skill./See *Correspondence*, V, 269, Appendix XXIX, for a fine 'free fall' demonstration of raillery of the second type: Swift to Lord Charlemont.

75 In a letter . . . among us'./*Correspondence*, V, 7.

Swift had a tendency to count heads./*Correspondence*, V, 270, Appendix XXX, 'Swift's Friends Classed by their Characters'.

For in your publick . . . to morrow./*Correspondence*, II, 44–5.

Defeat restores . . . isolation./For a fuller discussion of this theme see R. Paulson, 'Swift, Stella and Permanence', *ELH*, XXVII (1960), pp. 298–314.

Although critics . . . mature verse/e.g., H. Davis, *Jonathan Swift* (New York, 1964), 171–2. Davis's argument, originally published in 1931, that Temple's influence distorted Swift's 'natural bent' and that the Pindarics are a temporary diversion rather than a false start, is a suggestive one. It is necessarily qualified by later debate and research into the authorship of doubtful poems in the Christie Book and *The Whimsical Medley*.

Swift's attitude . . . ambiguous./As a 'practical politician' Swift took a shrewd view of the usefulness of anarchy; e.g., *Correspondence*, II, 113: 'I have been long afraid that we were losing the Rabble' (to Charles Ford, 7 August 1714); *Correspondence*, II, 131: 'If any thing witholds the Whigs from the utmost Violence, it will be onely the fear of provoking the Rabble . . .' (to Charles Ford, 27 September 1714).

76 so-called 'ungenerous' . . . invective/Editorial commentary, *Poems*, 296, 82. C. Peake, 'Swift's *Satirical Elegy on a Late Famous General*', *REL* III (1962), pp. 80–9,

vindicates Swift's method.

KEEPER . . . duly . . ./*Poems*, 835.

Sir Harold Williams . . . means./*Poems*, xvi.

77 Bless us, *Morgan* . . . Pity./*Poems*, 837.

The magisterial tone . . . pasturage-tithes./L. A. Landa, *Swift and the Church of Ireland* (Oxford, 1954), 135ff.

It has been said . . . contemporaries'/M. Johnson, *The Sin of Wit: Jonathan Swift as a Poet* (Syracuse, NY, 1950), 117.

Professor Ehrenpreis . . . Le Sage./Ehrenpreis, *Personality of Swift*, 43–8.

'On a F[art]Let in the House of Commons'/Formerly attributed to Prior, but rejected by Wright and Spears, *The Literary Works of Matthew Prior* (Oxford, 1959), II, 791. Apparently of mid-seventeenth century origin.

At last . . . his Ear.)/Prior, ed. cit., I, 264: 'Paulo Purganti and His Wife . . .'

78 My Lord . . . the Flames./'The Problem', 1699. *Poems*, 66.

He found . . . her Face./*Poems*, 589.

The catharsis . . . exclamation'./Johnson, op. cit., 112.

the real importance . . . *'mortal'*/See Kathleen Williams, *Jonathan Swift and the Age of Compromise*; paperback edition (Lawrence, Kansas, 1968), 148–53.

79 On Sense and Wit . . . round;/*Poems*, 593.

The Bason . . . spues./'The Lady's Dressing Room'. *Poems*, 527.

80 The Nymph . . . poison'd./'A Beautiful Young Nymph . . .' *Poems*, 583.

the complainants' . . . subjects./Ehrenpreis, *Personality of Swift*, 43.

Yet, some Devotion . . . a Thorn./*Poems*, 896.

81 It is a pity . . . criticisms of Swift./The phrase happens to be taken from Ricardo Quintana's humane and stimulating *Swift: an Introduction*, ed. cit., 142. I quote it quite out of context because this is what frequently occurs. It should be read here simply as a representative phrase. I am in no sense trying to short-circuit Professor Quintana's discussion.

Johnson remarks . . . affectation'./*Johnson: Prose and Poetry*, selected by M. Wilson, second edition (London, 1957), 319.

One recalls . . . 'the ruin to come'./Yeats, *Explorations*, 350.

Of 'more-pow'r' . . . woman'. See Johnson, *The Sin of Wit*, 26–7.

82 Yes, says the *Steward* . . . all to naught . . ./*Poems*, 71.

such as the anonymous 'South Sea Ballad' of 1720/See *Roxburghe Ballads*: VIII, part 1 (1897), 256.

Swift's Pindaric 'aberration'/Ball, op. cit., 16.

And so say I . . . be civil . . ./'Mary the Cook-Maid's Letter', 1718. *Poems*, 986.

83 He braced . . . possessed./Lecky, op. cit., II, 427.

And thus . . . Buffoonery./*Sensus Communis: an Essay on the Freedom of Wit and Humour* (1709), I, iv; see also W. K. Wimsatt (ed.), *English Stage Comedy* (New York, 1955), 5.

6. *Redeeming the Time*

84 The smaller . . . possibility./Iris Murdoch, *Sartre: Romantic Rationalist*, paperback reprint (London, 1961), 29.

Faced with such a statement . . . deceived itself./It will be objected that the French experience of the nineteenth century was in many ways different from that of British society. I agree. However, Iris Murdoch's discussion (pp. 26ff.) places the topic in a general European context of the nineteenth and twentieth centuries. Joyce, Conrad, Woolf and I. A. Richards are named together with Rimbaud, Mallarmé, Proust and Sartre.

Henry Adams . . . 'silence is best'/Adams, *The Degradation of the Democratic Dogma*, paperback edition (New York, 1958), 131.

Dr E. P. Thompson . . . denunciation'./Thompson, *The Making of the English Working Class*, paperback edition (Harmondsworth, 1968), 682.

R. H. Tawney . . . collapsed'./Tawney, Preface to *Life and Struggles of William Lovett*, 1876 (London, 1967), xiv.

85 Charles L. Stevenson . . . . attitudes'./Stevenson, *Ethics and Language*, paperback edition (New Haven and London, 1960), 210.

Theodore Redpath . . . Newton./Redpath, 'John Locke and the Rhetoric of the Second Treatise', in *The English Mind: Studies in the English Moralists Presented to Basil Willey*, ed. Hugh Sykes Davies and George Watson (Cambridge, 1964), 69.

Here I stand . . . THE BILL . . ./Quoted in Cecil Driver, *Tory Radical: The Life of Richard Oastler* (New York, 1946), 197.

'the striking change . . . doubt./ibid.

86 England has been described . . . society/F. Engels, *The Condition of the Working Class in England*, paperback edition (London, 1969), 37.

It is true . . . the winds./Quoted in Thompson, op. cit., 586.

87 Oastler is not 'compelling' but compelled./See Driver, op. cit., 197.

It has been pointed out . . . iambic'/George Saintsbury, *Historical Manual of English Prosody* (London, 1910), 200 n. 1.

'broken-backed'/C. C. Clarke, *Romantic Paradox: An Essay on the Poetry of Wordsworth* (London, 1962), 88.

There have been . . . embers') . . ./*The Correspondence of Gerard Manley Hopkins and Richard Watson Dixon*, ed. C. C. Abbott (London, 1935), 147–8.

88 Writers on linguistics . . . silence'./Seymour Chatman, in *The Kenyon Review*, 18 (Winter 1956), p. 422.

But Adam's thoughts . . . could have done . . ./George Eliot, *Adam Bede*, edited with an introduction by John Paterson (Boston, 1968), 170.

89 'the speech of the landscape'/George Eliot, *Impressions of Theophrastus Such*, third edition (Edinburgh and London, 1879), 48.

It is of course . . . language'./ibid., 44.

Such a style . . . England'./ibid., 38.

But I come back . . . people's side./*Essays of George Eliot*, ed. T. Pinney (London, 1963), 421–3.

An early critic . . . Tory-Democrat'./Joseph Jacobs, 1895, quoted in Pinney (ed.), op. cit., 415.

90 On 28 January 1810 . . . Thomas Poole./*Collected Letters of Samuel Taylor Coleridge*, ed. Earl Leslie Griggs, 6 vols (Oxford, 1956–71), III, 281.

'fixities . . . association'/*Biographia Literaria* by S. T. Coleridge, ed. J. Shawcross, 2 vols (Oxford, 1907), I, 202.

Of Parentheses . . . Hortus siccus./*Collected Letters of Coleridge*, III, 282.

For one . . . end'/Arthur Symons, introduction to *Biographia Literaria*, Everyman edition (London, 1906), xi.

L. C. Knights . . . humanity'/Knights, *Public Voices: Literature and Politics with Special Reference to the Seventeenth Century* (London, 1971), 24.

'moral *copula* . . . fatalism'/Coleridge Table Talk, 12 September 1831. *The Table Talk and Omniana of Samuel Taylor Coleridge*, with a note by Coventry Patmore (London, 1917), 157.

91 For if words . . . humanized./Coleridge, 'Author's Preface' to *Aids to Reflection*, new edition, revised (Edinburgh, 1896), xvii.

On some future . . . watch-words . . ./*Biographia Literaria*, ed. Shawcross, II, 116–17.

The architecture . . . economy./*Autobiography of Edward Gibbon as originally edited by Lord Sheffield*, World's Classics edition, reprint (London, 1950), 143–4.

By fencing . . . prejudice./ibid., 80.

'tendentious . . . image'/See Redpath, op. cit., 55, 69.

92 [Chesterfield . . . wit./Oliver Elton, *A Survey of English Literature: 1730–1780*, 2 vols (London, 1928), I, 9.

His definition . . . precedents./Table Talk, 12 July 1827; 9 May 1830; 31 May 1830. Ed. cit., 73, 91, 108.

'thro' . . . words . . . general'/*Collected Letters of Coleridge*, III, 281.

[Newman's . . . at all./J. M. Cameron, *The Night Battle: Essays* (London, 1962), 204.

These comments . . . Anglican period./ibid., 203.

The structure . . . undergoing./Walter E. Houghton, *The Art of Newman's 'Apologia'* (New Haven, 1945), 51–2. The two sentences under discussion are from pp. 260–1 of the 1913 edition of Newman's work.

John Beer . . . arguments'/Beer, 'Newman and the Romantic Sensibility', in Davies and Watson (ed.), *The English Mind*, 215.

93 I shall endeavor . . . done my best./*Collected Letters of Coleridge*, III, 282.

If I may dare . . . endure'./*Biographia Literaria*, ed. Shawcross, I, 65.

The metaphysical . . . common sense./Table Talk, 28 June 1834. Ed. cit., 311.

Mr Roebuck . . . themselves . . ./*The Complete Prose Works of Matthew Arnold*, III, *Lectures and Essays in Criticism*, ed. R. H. Super (Ann Arbor, 1962), 274.

94 has any one . . . poor thing!/ibid., 273.

In 1836 . . . Barraclough!'/See C. Aspin, *Lancashire: the first Industrial Society* (Helmshore, 1969), 117.

Plainly . . . intelligible./*The Letters of Gerard Manley Hopkins to Robert Bridges*, ed. C. C. Abbott (London, 1935), 265–6.

Hopkins wrote . . . business'/ibid., 284.

'jaded'/ibid., 201.

95 In 1832 . . . the reader'./Quoted by Noel Annan, 'John Stuart Mill', in Davies and Watson (ed.), *The English Mind*, 233.

men who speak . . . Sherry./*The Poetical Works of Wordsworth*, a new edition, revised by Ernest de Selincourt, reset (London, 1950), 737.

Mrs Transome . . . ring hollow./George Eliot, *Felix Holt, the Radical*, ed. F. C. Thomson (Oxford, 1980), 27.

It is, in fact . . . artisans./Andrew Ure, *The Philosophy of Manufactures or, an Exposition of the Scientific, Moral, and Commercial Economy of the Factory System of Great Britain* (London, 1835) 23, quoted in Thompson, op. cit., 396 n. 1.

96 I cannot express . . . moment./Quoted in F. M. Leventhal, *Respectable Radical: George Howell and Victorian Working Class Politics* (Cambridge, Mass., 1971), 11.

How strange . . . luxury./Huddersfield Committee pamphlet 'Humanity against Tyranny', quoted in Driver, op. cit., 106.

97 F. M. Leventhal . . . status./Leventhal, op. cit., 12.

Hopkins wrote . . . speech . . .'/Hopkins, *Letters to Bridges*, 46.

If asked . . . organic time'./Hans Meyerhoff, *Time in Literature* (Berkeley, Calif. 1955), 107; Daniel D. Pearlman, *The Barb of Time* (New York, 1969), 22.

98 It is as though . . . cannot./I consider my point to be justified even though the first 11½ lines of the 20-line poem concern working and belonging. What is the key-note of this belonging? – 'Little I reck ho!' (l. 9). Tom, in work, may be heedlessly happy but I cannot believe that Hopkins regarded this state as ideal. Little recking could so easily turn to great wrecking (cf. *Letters to Bridges*, 27–8). The verbal texture is lumpy, not sinewy.

In the contemporary . . . end came./Reprinted in the appendix to *Immortal Diamond*, ed. N. Weyand and R. V. Schoder (London, 1949).

Dom David Knowles . . . OUT./Knowles, *The English Mystical Tradition* (London, 1961), 97.

Dame Helen Gardner . . . nakid beyng of him'./Gardner, in *Essays and Studies*, XXII, (1937), p. 106.

99 Hopkins describes . . . always intoned./*Further Letters of Gerard Manley Hopkins including his Correspondence with Coventry Patmore*, ed. C. C. Abbott, second edition, revised and enlarged (London, 1956), 114.

Intonation . . . Aberdeenshire' 1791./*OED*.

'An "intonation pattern" . . . phrase'./Chatman, op. cit., p. 422.

'The opening phrase . . . choristers'./*OED*.

In 1859 . . . mankind'/*The George Eliot Letters*, ed. Gordon S. Haight, 9 vols (London and New Haven, 1954–78), III, 231. See also A. E. S. Viner, *George Eliot* (Edinburgh, 1971), p. 9.

100 The medieval . . . England . . ./Quoted in Alfred Thomas, SJ, *Hopkins the Jesuit* (London, 1969), 7 n. 1.

I am surprised . . . Catholicism./*Further Letters of Hopkins*, 93.

Michael Trappes-Lomax . . . stables'./Trappes-Lomax, *Pugin: A Medieval Victorian* (London, 1932), 55.

he complained that . . . Lincoln races'./Quoted in Denis Gwynn, *Lord Shrewsbury, Pugin and the Catholic Revival* (London, 1946), 15.

Father Walter J. Ong SJ . . . around him?/Ong, in Weyand and Schoder (ed.), *Immortal Diamond*, 100–101.

101 When D. P. McGuire . . . modern life'/Quoted in W. H. Gardner, *Gerard Manley Hopkins: A Study of Poetic Idiosyncrasy in Relation to Poetic Tradition*, 2 vols (London, 1944, 1949), II, 371.

'We lash . . . last!'/'The Wreck of the Deutschland', st. 8, ll. 2–3. *The Poems of Gerard Manley Hopkins*, fourth edition, ed. W. H. Gardner and N. H. MacKenzie (London, 1970), 54.

'christens her wild-worst Best'/'The Wreck of the Deutschland', st. 24, l. 8. ibid., 59.

the Corpus Christi procession/Hopkins, *Letters to Bridges*, 149: '. . . the procession . . . represents the process of the Incarnation and the world's redemption'.

In the death-cry . . . sacrifice./See E. Margaret Thompson, *The Carthusian Order in England* (London, 1930), 400.

The copy ... '*Martyrium*'./Evelyn Waugh, *Edmund Campion*, second edition (London, 1947), 55. Waugh attributes the French phrase to Canon Didiot.

102 'Extremes meet . . . Echo']./Hopkins, *Letters to Bridges*, 157.

'you cannot eat . . . difference'/ibid.

The only good . . . voice./ ibid., 280.

'recurb and . . . recovery'/'The Wreck of the Deutschland', st. 32, l. 3. Hopkins, *Poems*, ed. cit., 62.

103 Hopkins wrote . . . enjoyment./Donald Davie, 'Hopkins as a Decadent Critic', in *Purity of Diction in English Verse* (London, 1952), 171. I am aware that the *OED* allows, as early as 1631, the use of 'jaded' to suggest 'dull or sated by continual use or indulgence'. Matthew Arnold writes of his brother William who died, aged thirty-one, worn out with service in India, as 'jaded'. Milton, in the *Ludlow Mask* of 1634, has Comus, at his most speciously suasive, speak of delights 'to please, and sate the curious taste'. I suggest that Arnold and Milton use each word with particular justice, and that to turn one of Hopkins's words against himself, to make him appear like a Victorian Comus, is a little less than just.

7. *'Perplexed Persistence': The Exemplary Failure of T. H. Green*

104 'his value . . . self-contradictory'/*Works of Thomas Hill Green*, ed. R. L. Nettleship, 3 vols (London, 1888), III, 104.

'[he] sent . . . Truth'/B. M. G. Reardon, *From Coleridge to Gore: a Century of Religious Thought in Britain* (London, 1971), 305: quoting Henry Scott Holland.

'He would not . . . but it is true'/Green, *Works*, III, clxi.

'Abstract . . . nothing'/*Prolegomena to Ethics by . . . Thomas Hill Green*, ed. A. C. Bradley, fifth edition, reprinted (Oxford, 1929), 33.

'true philosophic . . . what he is'/*Works*, III, 116–17.

'content . . . principles'/ibid., 104.

'Man reads . . . things'/ibid.

'unconditionally . . . world'/Henry Sidgwick, *Outlines of the History of Ethics*, fifth edition (London, 1902), 272, 276. The phrases are from Sidgwick's paraphrase and discussion of Kant's philosophy.

105 the ineluctable fact/I wish to acknowledge my indebtedness, here and elsewhere in this essay, to D. M. MacKinnon's *A Study in Ethical Theory* (London, 1957).

Coleridge . . . Belief . . .'/Quoted in *Coleridge on the Seventeenth Century*, ed. R. F. Brinkley. Reprint (New York, 1968), 167–8.

In *Prolegomena* . . . as a whole'./*Prolegomena*, ed. cit., 81.

He further contends . . . political life'./ibid., 105.

'Thus the "Treatise . . . the new'./*Works*, I, 3.

'every explanation . . . metaphor'/ibid., III, 82.

106 to which Whitehead draws attention/A. N. Whitehead, *Symbolism* (Cambridge, 1928). The quoted phrases are from pp. 86–7.

'We hold . . . allies'/F. R. Leavis (ed.), *Mill on Bentham and Coleridge*, fifth impression (London, 1971), 140.

'call for . . . other classes'/Alfred Marshall, *Principles of Economics*, seventh edition (London, 1916), 45.

'the language of ethics' . . . despot'./MacKinnon, op. cit., 185, 65.

'the chasm . . . object'/J. H. Muirhead, *Coleridge as Philosopher* (London, 1930), 93.

'when every . . . language'/*Works*, III, 475.

'compound . . . healthy-mindedness'./Quoted in Melvin Richter, *The Politics of Conscience: T. H. Green and His Age* (London, 1964), 44.

107 He deplored, in *Prolegomena . . . chance'./Prolegomena*, 288.

'Green . . . with it'/H. J. Laski, 'The Decline of Liberalism', *L. T. Hobhouse Memorial Trust Lectures*, no. 10 (1940), p. 11.

'the best . . . grammar'/P. T. Forsyth, *Religion in Recent Art* (Manchester, 1889), 231.

'Oh, how I sympathise . . . criticism . . .'/A. and E. M. Sidgwick, *Henry Sidgwick: A Memoir* (London, 1906), 177.

108 'points beyond . . . formal implication'/M. H. Carré, *Phases of Thought in England* (Oxford, 1949), 368.

which Whitehead has so precisely described./Whitehead, op. cit., 88.

'"*Bona fide* . . . interest'./*Prolegomena*, 403 n. 1.

'Wordsworthian sentiment . . . principle'/Richter, op. cit., 47–8.

Green described . . . poetry'./*Works*, III, xviii.

In England . . . really./ibid., 118.

109 'The natural . . . man'./ibid., 119.

'the fixed . . . view'/A. C. Bradley, *Oxford Lectures on Poetry*. Reprint of the second edition (London, 1965), 131.

'that perplexed . . . question'/ibid., 136.

It is evident . . . do not./ibid., 131.

Leslie Stephen said . . . perplexed"./Stephen, *Studies of a Biographer*, new edition, 2 vols (London, 1929), I, 248.

110 'Yet . . . breath . . .'/Bradley, op. cit., 136–7.

The ethical . . . is./Hastings Rashdall, *The Theory of Good and Evil*, 2 vols (Oxford, 1907), I, 216–17. The work is dedicated 'To the memory of my teachers Thomas Hill Green and Henry Sidgwick'.

I had . . . melody./Quoted in Richter, op. cit., 162.

Richter says . . . inarticulate'./ibid., 80, 158.

'Few . . . said'/Green, *Works*, III, lxii.

'with singular clearness . . . time'/J. H. Muirhead, *Reflections*, ed. J. W. Harvey (London, 1942), 42.

'almost confounding humility,'/ibid.

I . . . written word./Quoted in Richter, op. cit., 14.

111 Coleridge commenced . . . harmonies'./*The Collected Works of S. T. Coleridge* (Bollingen Series LXXXV): *The Friend*, edited by Barbara E. Rooke, 2 vols (Princeton and London, 1969), I, 14.

'the charlatanry of common sense'/*Works*, I, 168.

'agreeable sensations and reflections'/ibid., III, 98.

'cultivated opinion' . . . 'confused'/ibid., lxxiv.

'wilfulness' and shallowness/ibid., xliv.

In *Prolegomena . . . thorough'./Prolegomena*, 168.

'shy . . . phrase'./*The Poetical Works of Wordsworth*, a new edition, revised by Ernest de Selincourt, reset (London, 1950), 582.

'the oral channel . . . worked'/H. Sykes Davies and G. Watson (ed.), *The English Mind: Studies in the English Moralists presented to Basil Willey* (Cambridge, 1964), 183.

112 'valuable warnings . . . sins'/Richard Chenevix Trench, *On the Study of Words*, sixth edition (London, 1855), 42.

'how deep . . . many words'/Trench, loc. cit. I am indebted to Austin Warren, *Rage for Order* (Chicago, 1948), 61, 63; to Hugh Kenner, *The Pound Era*, new edition (London, 1975), 102–5; and to an essay by Darcy O'Brien in *New Blackfriars* (October 1972), pp. 466–70.

'there is no alternative . . . their way'/*Prolegomena*, 419.

'It may . . . resisting'./ibid., 117.

'Success . . . before him'./Quoted in Richter, op. cit., 150.

'Those . . . *tilth*'/Green, *Works*, III, xviii.

' "Swing . . . in it" './ibid., xxxvii–xxxviii.

'came in with Kingsley . . . blasted nonsense'/*The Correspondence of Gerard Manley Hopkins and Richard Watson Dixon*, ed. C. C. Abbott (London, 1935), 74.

113 had a theory . . . syllogism./*Works*, III, xxxviii.

'locked . . . case'/Elisabeth W. Schneider, *The Dragon in the Gate: Studies in the Poetry of G. M. Hopkins* (Berkeley, Calif., 1968), 35. Subsequent quotations are from pp. 35–6.

'all the *hooks* . . . metaphor'/Coleridge, *The Friend*, ed. cit., I, 20–1.

'earnestness' . . . energy'/*Works*, III, 353–4.

Though he had . . . all./ibid., lxiv–lxv.

114 'unconscious social insolence'/ibid., 460.

'assumption of superiority'/Coleridge, *The Friend*, II, 277.

'the characteristic . . . experience'./A. J. M. Milne, *The Social Philosophy of English Idealism* (London, 1962), 89.

'weakness . . . filled up'/*Works*, III, 91.

'signs of the strain . . . values'/Richter, op. cit., 201.

Newman wrote . . . arguments'./Newman, *An Essay in Aid of a Grammar of Assent* (London, 1870), 408.

'against mere impulse . . . alike'/A. and E. M. Sidgwick, op. cit., 243. (H. S. to Roden Noel, 1871).

Sidgwick's 'law . . . flexible'/ibid.

'to enlighten . . . meditation'/Leavis (ed.), op. cit., 99.

115 'like a good rider . . . horse'/Walter Bagehot, *Literary Studies*, ed. R. H. Hutton, 2 vols (London, 1879), I, 171.

'the dogmatic . . . *dichotomy*'/Muirhead, *Coleridge*, 83.

as Richter has claimed/Richter, op. cit., 47.

'unfused . . . conspicuous'/*Works*, III, 88.

'The account . . . actualised" './ibid., 76.

'As the poet . . . place . . .'/ibid., 90.

'how the moral intelligence gets into poetry'/Allen Tate, *Essays of Four Decades* (London, 1970), 149, quoting and discussing a phrase of Yvor Winters.

'The new students . . . items appear'/Carré, op. cit., 367.

116 'the form . . . data'/ibid., 368.

Not as to the sensual . . . sign./Green, *Works*, III, 273.

Shakespeare, Keats and Tennyson/e.g. III, 90, 273–4.

'the indestructible . . . life'/A and E. M. Sidgwick, op. cit., 541: referring to section cxxiv of the poem.

A pioneering study . . . existentially" './A. Birch Hoyle, *The Teaching of Karl Barth*

(London, 1930), 226. Cf. A. and E. M. Sidgwick, op. cit., 539–41; D. G. James, *Henry Sidgwick: Science and Faith in Victorian England* (London, 1970), 39–41.

She ceased . . . lay/'The Female Vagrant', *Wordsworth: Poetry & Prose*, selected by W. M. Merchant (London, 1969), 148.

117   the consciousness . . . found it./*Philosophical Lectures and Remains of Richard Lewis Nettleship*, ed. A. C. Bradley, second edition (London, 1901), 132.

'approach philosophy . . . language'/ibid., li–lii n. 2.

'the "I" . . . feeling'/*Works*, III, 104.

As has been . . . transmute them./ibid., 72.

For Green . . . developed'./ibid.

118   'imperfect . . . progressive'/ibid., 73.

'implies . . . self'/ibid., 72.

'[Green] . . . self-sacrifice'/Holland, quoted in Reardon, op. cit., 305 n. 2.

He may 'find . . . repentance./*Prolegomena*, 108.

'to the significant . . . imagination'/Dorothea Krook, *Three Traditions of Moral Thought* (Cambridge, 1959), 10.

'From Coleridge . . . lay writers'./W. R. Inge, *The Platonic Tradition in English Religious Thought* (London, 1926), 95.

'deep seriousness . . . place'/ibid., 90.

'when the poet . . . elements'/*Works*, III, 45.

his own argument/ibid., 104–5.

119   'I could have done . . . did something./MacKinnon, op. cit., 129.

'An Answer to Mr. Hodgson',/*Works*, I, 521–41.

The original subscribers . . . four hundred./Coleridge, *The Friend*, ed. cit., II, 407.

'to direct his remarks . . . clerisy".'/J. R. de J. Jackson, *Method and Imagination in Coleridge's Criticism* (London, 1969), 32.

'the attention . . . fellow-labourer'/Coleridge, *The Friend*, I, 21.

'abstruse and laboured'/ibid., II, 442.

'A man . . . engaged'./Peter Brook, *The Empty Space*, paperback edition (Harmondsworth, 1972), 11.

120   'passed through . . . rest'/Forsyth, op. cit., 189. Forsyth is not here discussing Green, though he knew Green's work. See J. H. Rodgers, *The Theology of P. T. Forsyth* (London, 1965), 312.

8. *What Devil Has Got Into John Ransom?*

121   The question . . . contentious/*The Literary Correspondence of Donald Davidson and Allen Tate* ed. J. T. Fain and T. D. Young (Athens, Georgia, 1974), 344.

'It will seem . . . unfriendship'/Quoted in Michael O'Brien, *The Idea of the American South 1920–1941* (Baltimore and London, 1979), 134.

To Allen Tate . . . middle age./Allen Tate, *Memories & Essays . . . 1926–1974* (Manchester, 1976), 43.

'meaty and flavorsome'/*Davidson-Tate Correspondence*, 284

'John . . . indeed'/ibid., 104.

'cruelly polite snobbishness'/ibid., 322.

'if . . . art'/John Crowe Ransom, *The World's Body*, paperback edition (Baton Rouge, 1968), 211.

'it is common . . . argument'/Ransom, *The New Criticism* (Norfolk, Conn., 1941), 295.

'argument . . . argument'/ibid.

122   'a good poem' . . . self'/*The World's Body*, 2.

'vocalism' . . . prose'/Ransom, *Beating the Bushes: Selected Essays 1941–1970* (New York, 1972), 175–6.

'nothing . . . manoeuvre'/*The World's Body*, 347.

Allen Tate declared/Quoted in T. D. Young, *Gentleman in a Dustcoat: A Biography of John Crowe Ransom* (Baton Rouge, 1976), 465.

Ransom's definition/*The World's Body*, 348.

'wit . . . detachment'/ibid., 3.

'torture of equilibrium'/'The Equilibrists': John Crowe Ransom, *Selected Poems*, third edition revised and enlarged (London, 1970), 85.

'the supreme equilibrist'/Miller Williams, *The Poetry of John Crowe Ransom* (New Brunswick, NJ, 1972), 20.

'balance . . . poetry'/ibid., 11.

'Eclogue'/*Selected Poems*, ed. cit., 15.

'Survey of Literature'/ibid., 68–9.

123   'nervous strain . . . moderns'/*The New Criticism*, 38.

'an intolerable . . . surface'/Tate, *Memories & Essays*, 42.

'an intense sympathy' . . . serenity'/Young, *Gentleman in a Dustcoat*, 460.

George Eliot's dictum . . . life'/George Eliot, *Felix Holt, the Radical* (1866), ed. P. Coveney (Harmondsworth, 1972), 129.

'an act of will' . . . 'attention'/*The World's Body*, 236; *The New Criticism*, 273.

'The plight . . . desperate'/*The World's Body*, 253.

'I am an Anglophile . . . American'/Ransom, *God Without Thunder: An Unorthodox Defence of Orthodoxy*, UK edition (London, 1931), 356.

'John . . . mystery'/*Davidson-Tate Correspondence*, 312.

'respect . . . circumference'/*Selected Poems*, ed. cit., 59.

124   'Far . . . center'/ibid., 37.

'It . . . nature'/ibid., 66.

'workmanlike . . . argument'/ibid., 152.

'metaphysical . . . eccentric'/*The New Criticism*, 190. cf. Tate's pejorative use of 'eccentric', *Davidson-Tate Correspondence*, 140.

'middling ways'/*Selected Poems*, ed. cit., 58.

'the three careers . . . center'/Robert Buffington, *The Equilibrist: A Study of John Crowe Ransom's Poems, 1916–1963* (Nashville, Tenn., 1967), 2.

'keep . . . separate'/ibid., 1.

'the agony of composition'/*The World's Body*, 348.

'reliance . . . judgment'/*Memories & Essays*, 43.

'calm . . . remorseless'/Louis D. Rubin, Jnr, *The Wary Fugitives: Four Poets and the South* (Baton Rouge & London, 1978), 14, 49, 55,

the more literal mind/but cf. Young, *Gentleman in a Dustcoat*, 140.

'eccentricities . . . thesis'/Graham Hough, in *John Crowe Ransom: Critical Essays and a Bibliography* ed. T. D. Young (Baton Rouge, 1968), 187–9.

125   'no art . . . faculties'/Quoted in Rubin, *Wary Fugitives*, 49.

'the rational power . . . Spectre'./S. Foster Damon, *A Blake Dictionary*, first UK edition (London, 1973), 380–2.

logic of 'abstraction'/Cf. J. M. Bradbury, *The Fugitives: A Critical Account* (Chapel Hill, 1958), 129.
'logical structure' . . . 'beautiful poem'/*The New Criticism*, 53.
'natural . . . judgments'/Young, *Gentleman in a Dustcoat*, 437.
'meditative . . . effect/*Beating the Bushes*, 175–6.
'remorseless' . . . 'ruthless'/Rubin, *Wary Fugitives*, 55–6; Ransom, *God Without Thunder*, 127.
'we may hardly . . . everybody'/*The World's Body*, 279.
'the kind of poetry . . . adult mind'/ibid., viii.
'Images . . . darkness'/ibid., 116–17.

126   took Kant as his mentor/Young, *Gentleman in a Dustcoat*, 423.
'perpetually disquieted'/*God Without Thunder*, 304.
'a man . . . "situation"'/*Beating the Bushes*, 82.
Kant . . . mutilation'/*God Without Thunder*, 304.
The phrase is William Empson's/Empson, 'This Last Pain', *Collected Poems* (London, 1955), 33.
The Kenyon Critics/ . . . *Studies in Modern Literature from the Kenyon Review* . . . (Cleveland and New York, 1951).
Ransom published . . ./Young (ed.), *John Crowe Ransom* . . . , 243.
Christopher Norris/*William Empson and the Philosophy of Literary Criticism* (London, 1978), 124, 213 n. 31.
'gestures not of deference'/*Selected Poems*, ed. cit., 59.

127   'Our Two Worthies'/ibid., 61.
'obstinate intellectual waywardness'/Graham Hough, in Young (ed.), *John Crowe Ransom* . . . , 189.
'beautiful thing'/Young, *Gentleman in a Dustcoat*, 478.
'academic standing' . . . meat'/*Davidson-Tate Correspondence*, 87.
'seriously concerned . . . efforts'/Young, *Gentleman in a Dustcoat*, 140.
Rubin's suggestion/Rubin, *Wary Fugitives*, 280.
His hypothesis . . . modernity'/*The World's Body*, 11–12.
amiable unpretentiousness . . . gentleness'/Rubin, *Wary Fugitives*, 63; Young, *Gentleman in a Dustcoat*, 326.
'self-expression' . . . feathered'/*Davidson-Tate Correspondence*, 172.
'two kinds . . . philosophical'/*The World's Body*, 34.

128   Schopenhauer's 'knowledge without desire'/ibid., 45.
'predatory . . . interest'/ibid., 38.
'pitch of attention'/*The New Criticism*, 273.
'the groan . . . sighs'/Robert Graves, *Collected Poems: 1914–1947* (London, 1948), 23: 'Lost Love'.
As Vivienne Koch . . . structure./Koch, in Young (ed.), *John Crowe Ransom* . . . , 127.
'Eclogue'/For information on the successive versions of this poem I am indebted to Robert Buffington, op. cit., 135–6.
'disparity' . . . 'real logic'/*The New Criticism*, 242.

129   F. O. Matthiessen suggested/Matthiessen, in Young (ed.), *John Crowe Ransom* . . . , 85.
'consider . . . conscience'/*The World's Body*, 91.
'the confusion . . . density'/*The New Criticism*, 79.

'Platonic . . . monsters'/*The World's Body*, 225.
130 'good poet'/*The New Criticism*, 295.
'good poem'/*The World's Body*, 2.
'beautiful poem'/*The New Criticism*, 53.
'an absorbing . . . mania'/*The World's Body*, 72.
'a simplified . . . mania'/*The New Criticism*, 165.
'great . . . there'/Davidson, quoted in Young, *Gentleman in a Dustcoat*, 123.
'obligations' . . . authorship'/*The World's Body*, xv.
'In this century . . . story'/Rubin, *Wary Fugitives*, 356.
'naturalistic rabble'/*God Without Thunder*, 349.
'aggressions'/*The World's Body*, 198 n. 1.
'cold fury'/*Beating the Bushes*, 112.
131 'Poetic . . . technique'/*The New Criticism*, 79.
'ritual . . . occasions'/ibid., 205.
'It is not . . . thing'./*The World's Body*, 59.
Allen Tate/In *Essays of Four Decades* (London, 1970), 466.
'kind of obliquity'/*The World's Body*, 32.
'We are creatures . . . features'/*God Without Thunder*, 303.
Modern poetry, Tate said/*Davidson-Tate Correspondence*, 140.
'shaken . . . not as a leaf'/*Selected Poems*, ed. cit., 125.
Ransom's critics/e.g. Young, *Gentleman in a Dustcoat*, 375–7.
'objectionable . . . staple'/*The New Criticism*, 253.
'were agreed . . . evidences'/ibid., 99.
'nature still waits . . . contingent'/*God Without Thunder*, 153.
'things . . . materiality'/*The World's Body*, 116.
'the world . . . objects'/ibid., 123.
'the vast . . . contingency'/*Beating the Bushes*, 107.
'the stubborn . . . world'/*God Without Thunder*, 255.
132 'marvelling . . . substance'/*The World's Body*, 142.
'the denser . . . world'/*The New Criticism*, 281.
'ontological . . . poets'/ibid., 335.
'stubborn substance'/*Beating the Bushes*, 61.
'dense natural context'/ibid., 109.
F. P. Jarvis/In Young (ed.), *John Crowe Ransom . . .* , 206–9.
Bradley embodied . . . fact'/A. C. Bradley, *Oxford Lectures on Poetry*, reprint of the second edition (London, 1965), 71.
in a late essay/*Beating the Bushes*, 172.
'self-division . . . good'/Bradley, loc. cit.
Ransom argues here/Quoted in Young, *Gentleman in a Dustcoat*, 89.
'[Poetry's] nature . . . autonomous'/Bradley, op. cit., 5.
'the artist . . . not real'/*The World's Body*, 198.
'As science . . . body'/ibid.
'sly analogy' . . . language'/*The New Criticism*, 79.
133 'A poem . . . world'/*Beating the Bushes*, 75.
Ransom once . . . metaphors'/*The World's Body*, 137.
in one of his last essays/*Beating the Bushes*, 174.
'majesty . . . contingency'/*God Without Thunder*, 333–4.
'Figures . . . course'/*The World's Body*, 133.

'Why . . . were'/*Selected Poems*, ed. cit., 4.
'And . . . Remember'/ibid., 18.
'No . . . fear'/ibid., 34.
'A cry . . . heart'/ibid., 37.
'Honor . . . doves'/ibid., 85.
'Whose ministers . . . saith"'/ibid., 101.
'until . . . again'/ibid., 111.
'Conrad . . . Conrad!'/ibid., 118.
Captain Carpenter/ibid., 44–6.
134 In an essay of 1942/*Beating the Bushes*, 70–1.
'monad' . . . existence'/*OED*.
as the medieval mystics perceived/*The Cloud of Unknowing*, ed. Dom Justin
McCann (London, 1947), xvii.
the passage from Hamlet . . . analysis/*Beating the Bushes*, 136–43.
'realism' . . . the fact'/ibid., 48.
135 Keats . . . in February 1818 . . . pulses/*The Letters of John Keats*, ed. M. Buxton
Forman, third edition (London, 1947), 108, 142.
'The rhythm . . . thought'/C. H. Sisson, *English Poetry 1900–1950: An Assessment*
(London, 1971), 30.
'the social issue . . . fiction'/*Selected Poems*, ed. cit., 148.
Robert Buffington/*The Equilibrist*, 112.
Raymond Williams's acute analysis/In *The Manchester Guardian*, 9 Dec. 1960,
quoted by Fr Martin Jarrett-Kerr, CR, 'The 491 Pitfalls of the Christian Artist', in
The Climate of Faith in Modern Literature ed. Nathan A. Scott, Jnr (New York,
1964), 203.
136 'behind . . . language'/*The New Criticism*, 79.
'the whole body . . . race'/*OED*.
'dull readers . . . affiliations'/*The New Criticism*, 102–3.
'those Yale . . . reader'/*Beating the Bushes*, 171, 176.
'that terrible . . . strategy'/*The World's Body*, 272.
'the last . . . poets'/*The New Criticism*, 275.
'the possibility . . . destruction'/*The World's Body*, 241.
'are easy . . . for them'/*Beating the Bushes*, 160.
'more intense . . . known'/Ransom, quoted in Young, *Gentleman in a Dustcoat*, 402.
'The bad artists . . . strangely possessed'/*The World's Body*, 39.
137 'humble reader'/*The New Criticism*, 158.
'plain reader' . . . pain'/Laura Riding and Robert Graves, *A Survey of Modernist
Poetry*, 2nd imp. (London, 1929), 106.

9. *Our Word Is Our Bond*

138 *And I might mention . . . concerned.*/J. L. Austin, P[hilosophical] P[apers] (Oxford,
1961), 227–8.
'Now for the *Poet* . . . lieth.'/*The Prose Works of Sir Philip Sidney*, ed. Albert
Feuillerat, 4 vols (Cambridge, 1912, reprinted 1962), III, 29.
'The Imagination . . . Truth.'/*The Letters of John Keats*, ed. Maurice Buxton
Forman, third edition (London, 1947), 68: To Benjamin Bailey, Saturday 22 Nov.
1817.

'The poem . . . about it.'/*The Collected Poems of Wallace Stevens* (London, 1955), 473: 'An Ordinary Evening in New Haven' XII.

'All values . . . judicial sentences.'/*The Letters of Ezra Pound 1907–1941*, ed. D. D. Paige (London, 1951), 249: To Felix E. Schelling, Paris, 8 July 1922.

intellectus agens/Dom Cuthbert Butler, *Western Mysticism*, third edition (London, 1960), 101 n. 2: 'Intellectus agens dicitur qui facit intelligibilia in potentia esse in actu . . .'; *The Sermons and Devotional Writings of Gerard Manley Hopkins*, ed. Christopher Devlin, SJ (London, 1959), 125: 'Here we touch the *intellectus agens* of the Averrhoists and the doctrine of the Hegelians and others.'

recta ratio factibilium/David Levy, 'Faith and Sensibility', in *P. N. Review*, edited by Donald Davie, Michael Schmidt and C. H. Sisson, vol. 4, no. 3, p. 63: 'The scholastics held that "art is a virtue of the practical intellect, *recta ratio factibilium*".'

mens rea and *actus reus*/John R. Silber, 'Being and Doing: A study of Status Responsibility and Voluntary Responsibility', in *The Anatomy of Knowledge*, ed. Marjorie Grene (London, 1969), 166: '*mens rea*, the awareness of the wrongfulness or unlawfulness of the conduct, and *actus reus*, the physical manifestation of *mens rea*'.

'boasting . . . carried'/Vernon Watkins, *Ballad of the Mari Llwyd and other poems*, second edition (London, 1947), 89.

139 'If you are a judge . . . a state of mind.'/J. L. Austin, [*How To Do Things With*] *Words*, ed. J. O. Urmson, 'reprinted from corrected sheets of the 1963 reprint' (Oxford, 1965), 88.

'A performative utterance . . . soliloquy'/ibid., 22.

'ordinary language' . . . practical matter'/Austin, *PP*, 130.

'parasitic . . . medium'/Austin, *Words*, 104.

'determined . . . prose'/Isaiah Berlin, *Personal Impressions* (London, 1980), 101–2.

'accuracy . . . *bond*'/Austin, *Words*, 10.

'controlled interplay'/Karl Britton, 'Symbolic Actions and Objects: "The weak pipe and the little drum", in *Philosophy*, 54 (1979), p. 289: 'The poem says something that arises from the controlled interplay of just these words.'

Austin's belief . . . 'philosophically unimportant'./Austin, *Words*, 5 n. 2.

'there appeared . . . exposition'/Berlin, op. cit., 104.

'the complex and recalcitrant nature of things'/ibid., 113.

'even "ordinary language" . . . extinct theories'/Austin, *PP*, 130.

'considerations . . . may infect statements'/Austin, *Words*, 55.

'perfectly neat and easy'/*Works of Thomas Hill Green*, ed. R. L. Nettleship, 3 vols (London, 1888), III, 108–9.

'distinguished . . . ideas'/*Berkeley's Commonplace Book*, ed. G. A. Johnston (London, 1930), 123 n. 156.

140 'manifold' and 'inextricable'/*Selections from Berkeley*, [edited] by A. C. Fraser, fifth edition, amended (Oxford, 1899), 23: 'Introduction to the Principles' 17.

'inflexible natures . . . of things'/*Locke's Conduct of the Understanding*, ed. Thomas Fowler, fifth edition (Oxford, 1901), 51: 'For to have right conceptions about them, we must bring our understandings to the inflexible natures and unalterable relations of things, and not endeavour to bring things to any preconceived notions of our own.'

'things themselves' . . . understood'/ibid.

'of excellent use' . . . 'impose on the understanding'/Fraser (ed.), *Selections from Berkeley*, 28: 'Introduction to the Principles' 21.

powers/I allude to the Coleridgean sense of the word. See S. T. Coleridge, 'Author's Preface' to *Aids to Reflection*, new edition revised (Edinburgh, 1896), xvii.

'all controversies . . . 'legitimate concepts'/Fraser (ed.), *Selections from Berkeley*, 28: 'Introduction to the Principles' 22, 29 n. 2.

'If it were not for Sin . . . as *Angels* do'/*Moral and Religious Aphorisms by Benjamin Whichcote DD* (London, 1930), 82.

'naked, undisguised ideas' 'stripped of words'/Fraser (ed.), *Selections from Berkeley*, 31: last words of 'Introduction to the Principles'; Fowler (ed.), *Locke's Conduct of the Understanding*, 92.

'controversial relish'/George Santayana, *Some Turns of Thought in Modern Philosophy: Five Essays* (Cambridge, 1933), 26.

'a jejune and dry way . . . mathematicians only'/Fowler (ed.), *Locke's Conduct of the Understanding*, 91.

'the short jejune way . . . ethiques'/Johnston (ed.), *Berkeley's Commonplace Book*, 17.

'plain unsophisticated arguments' . . . 'plausible discourses'/Fowler (ed.), *Locke's Conduct of the Understanding*, 91–2.

'proper sentiments' . . . 'ordinary affairs of life'/Donald Davie, *Articulate Energy: An Inquiry into the Syntax of English Poetry* (London, 1955), 121, quoting Berkeley: 'In the ordinary affairs of life, any phrases may be retain'd, so long as they excite in us proper sentiments . . .'

'signs or counters', 'fiduciary symbols'/Davie, loc. cit., quoting Berkeley, and St.-John Perse.

141 carefully-elaborated discourse/Frederick J. Powicke, *The Cambridge Platonists: A Study* (London, 1926), 51–2: '[Whichcote] seems to have introduced a new style. Instead of reading a carefully-elaborated discourse, he spoke from a few notes fluently, easily, and sometimes colloquially.'

Whichcote's 'candor'/ibid., 78, quoting Whichcote: 'God expects that the reader of Scripture should be of an ingenuous spirit, and use candor, and not lie at the catch.'

Hobbes's 'Perspicuous Words'/*Leviathan*, ed. C. B. Macpherson (Harmondsworth, 1968), 116: 'The Light of humane minds is Perspicuous Words, but by exact definitions first snuffed, and purged from ambiguity.'

'speak accurately'/Powicke, op. cit., 73.

Hobbes's audacity/Dorothea Krook, *Three Traditions of Moral Thought* (Cambridge, 1959), 106: 'Hobbes adds (with an audacity that Hooker would have found stupefying) . . .'

'troublesome multiplicity'/Powicke, op. cit., 74, quoting Whichcote: 'And this is fit for you to know to avoid a troublesome multiplicity in religion . . .'

the world's ineluctable necessity/Krook, op. cit., 95: 'The logic of the argument is inescapable . . . for it follows by an ineluctable necessity from [Hobbes's] view of man's nature.'

'real language of men' . . . dignity of man'/*The Poetical Works of Wordsworth*, a new edition revised by Ernest de Selincourt (London, 1950), 740, 737, 'Preface to the Second Edition of Several of the Foregoing Poems'.

'airy useless notions' . . . 'real and substantial knowledge'/Fowler (ed.), *Locke's Conduct of the Understanding*, 94.

another Lockean idiom/ibid., 95: 'To accustom ourselves in any question proposed, to examine and find out upon what it bottoms.'

'the wisdom of the world'/Green, *Works*, III, 233: 'The wisdom of the world . . . represents the mental state of what St Paul calls the carnal or natural man.'

'a sharpened awareness . . . phenomena'/Austin, *PP*, 130.

'patient accumulation of data about actual usage'/Berlin, op. cit., 113.

'the conduct of meetings and business'/Austin, *Words*, 156.

His work . . . H. L. Hart')./John Passmore, *A Hundred Years of Philosophy*, third edition (Harmondsworth, 1968), 598.

He requires . . . ideal ones'/Austin, *PP*, 38.

'mind is incorrigibly poetical' . . . 'many-coloured ideas'/Santayana, op. cit., 22–3.

'taking the sentences . . . the next one'/G. J. Warnock, in Isaiah Berlin and others, *Essays on J. L. Austin* (Oxford, 1973), 36.

'*in a peculiar way* hollow or void'/Austin, *Words*, 22.

142 Avoid the practice . . . other writers./*MHRA Style Book: Notes for Authors and Editors*, ed. A. S. Maney and R. L. Smallwood (Leeds, 1971), 16.

'"Wherefrom father Ennius . . . mould in plaster'/*Personae: Collected Shorter Poems of Ezra Pound* (London, 1952), 219, 198.

'egocentric naïveties' and 'obtuse assurance'/F. R. Leavis, *The Great Tradition* (London, 1948), 210: '[Conrad's] irony bears on the egocentric naïveties of moral conviction . . . and the obtuse assurance with which habit and self-interest assert absolute rights and wrongs.'

'compacted doctrines'/William Empson, *The Structure of Complex Words* (London, 1951), 39: 'it is often said . . . that a word can become a "compacted doctrine" . . .'

'philosophers . . . to any word'/J. L. Austin, *Sense and Sensibilia, reconstructed from the manuscript notes by G. J. Warnock* (Oxford, 1962), 62.

'the innumerable . . . language'/ibid., 73.

Charlotte Brontë retorted . . ./Quoted in Robert Bernard Martin, *The Accents of Persuasion* (London, 1966), 27.

'turned and picked up . . . Salient'/John Brophy and Eric Partridge (ed.), *Songs and Slang of the British Soldier 1914–1918*, second edition (London, 1930), 193–4.

'our chains rattle . . .'/*Biographia Literaria* by S. T. Coleridge, ed. J. Shawcross, 2 vols (Oxford, 1907), II, 116.

143 'If said by an actor . . . soliloquy'/Austin, *Words*, 22.

'surveying . . . ethical space'/ibid., 10.

'noumena' . . . 'nous' . . . 'sense or meaning'/*Coleridge on the Seventeenth Century*, ed. Roberta Florence Brinkley, reprint edition (New York, 1968), 145–6: Note on Hooker's *Eccl. Polity*, Book I, sec. 2, 'That which doth assign unto each thing the kind, that which doth moderate the force and power, that which doth appoint the form and measure of working, the same we term a *Law*.' STC observes: 'The law is Res *noumenon*; the Thing is Res *phoenomenon*.' See also H. G. Liddell and R. Scott, *Greek English Lexicon*, fifth edition revised and augmented (Oxford, 1864), 952.

'it is . . . a Poet by'/Sidney, *Prose Works*, III, 11.

This ioyous day . . . which the Lord vs taught./*The Poetical Works of Edmund Spenser*, ed. J. C. Smith and E. de Selincourt (London, 1912), 573.

144 'ye are dearely bought'/1 Corinthians 6: 20, in the 'Bishops' Bible', 1568: 'For ye are dearely bought: therefore glorifie God in your body and in your spirite, which are Gods.'

'felicissime audax' . . . 'curiosa felicitas'/*The Oxford Companion to Classical Litera-ture*, compiled and edited by Sir Paul Harvey, reprinted with corrections (Oxford, 1940), 215: 'Quintilian calls [Horace] "felicissime audax", and Petronius refers to his "curiosa felicitas" or "studied felicity".'

'extra-linguistic'/G. J. Warnock, in Berlin and others, *Essays* . . . , 73.

'contemplation' . . . 'beholding the Idaea playne'/*Amoretti* LXXX, LXXXVIII, *Poetical Works*, ed. cit., 575, 577.

'haste and botching'/J. W. Lever, *The Elizabethan Love Sonnet*, second edition (London, 1966) 100–101: 'Whatever Spenser's intentions were, there are unmis-takable signs of haste and botching in the little 1595 volume containing *Amoretti* and *Epithalamion.*'

'erected wit' . . . reaching unto it'/Sidney, *Prose Works*, III, 9.

Theodore Spencer . . . the other'/Theodore Spencer, 'The Poetry of Sir Philip Sidney', in *ELH*, XII (1945), p. 253.

'*Idea*' . . . things'/Sidney, *Prose Works*, III, 8, 13–14.

'captived . . . world'/ibid., 18.

'For my owne part . . . Wit'/*Poems and Dramas of Fulke Greville*, ed. Geoffrey Bullough, 2 vols (Edinburgh, n.d.), I, 13.

145 'But, ah, desire still cries . . .'/Sir Philip Sidney, *Selected Poems*, ed. Katherine Duncan-Jones (Oxford, 1973), 153.

'an erection of those engendring parts'/*OED*, citing Hugh Plat, *The Jewell House of Art and Nature*, 1594.

'dark and disputed matter'/Hopkins, *Sermons* . . . , ed. Devlin, 150.

problems which the writer 'unconsciously' raises/ibid., 339: 'the problems which GMH, sometimes unconsciously, raises'.

'bearing a part in the conversation'/Fowler (ed.), *Locke's Conduct of the Understand-ing*, 98: 'The shame that such dumps cause to well-bred people, when it carries them away from the company, where they should bear a part in the conversa-tion . . .'

The theologian D. M. MacKinnon/MacKinnon, *The Problem of Metaphysics* (Cambridge, 1974), 22, 39.

'there is that . . . what they are'/Green, *Works*, III, 87.

'primitive purity and shortness'/quoted in *Critical Essays of the Seventeenth Century*, ed. J. E. Spingarn, 3 vols (Oxford, 1908–9), II, 117–18.

'we can know . . . perversely'/Austin, *PP*, 142.

'providence of wit'/Dryden, 'To my Honored Friend Sir Robert Howard on his Excellent Poems', l. 34. *The Poems of John Dryden*, ed. James Kinsley, 4 vols (Oxford, 1958), I, 14.

'concupiscence of witt'/Donne, 'The Crosse', l. 58. John Donne, *The Divine Poems*, ed. Helen Gardner (Oxford, 1952), 26–8

'anarchy of witt'/Dryden, 'Prologue to Albumazar', l. 17. *Poems*, ed. cit., I. 141.

Cudworth . . . acted with . . .'/*A Sermon Preached Before the House of Commons, March 31, 1647*, Reproduced from the Original Edition (New York, 1930), 27.

146 Vaughan's 'But living . . . myre . . .'/*The Works of Henry Vaughan*, ed. L. C. Martin, 2 vols (Oxford, 1914), II, 522–3: 'The Night'.

'for if he . . . runne'./*Leviathan*, ed. cit., 194.

'doctrine of the *Infelicities*'/Austin, *Words*, 23; *PP*, 226–7.

'not everyone . . . I did'/Warnock, in Berlin and others, *Essays* . . . , 44.

192

'liked authority'... position'/ibid., 33; Berlin, *Personal Impressions* 105.

'learn from the distinctions encapsulated in their ordinary uses'/Warnock, in Berlin and others, *Essays*..., 37.

147 to drive 'with care'/G. W. Pitcher, ibid., 19.

'clumsily... clumsily'/Austin, *PP*, 147.

'without effect'... effects'/Austin, *Words*, 17.

jurists' 'timorous fiction... fact'/ibid., 4 n. 2.

'In these discussions... Austin'/Pitcher, in Berlin and others, *Essays*..., 21.

'magic circle'/Berlin, op. cit., 115.

'Elegance... 'Obscurity'/David Hume, autobiographical letter of 1734, quoted by Raymond Williams, 'David Hume: Reasoning and Experience', in *The English Mind: Studies in the English Moralists presented to Basil Willey*, ed. H. Sykes Davies and G. Watson (Cambridge, 1964), 125.

'most excelle[n]t... learning'/Sidney, *Prose Works*, III, 21.

Hopkins's ambiguous... pieces...')/Dark and disputed matter, to my mind. Compare, for example, *The Poems of Gerard Manley Hopkins*, fourth edition, ed. W. H. Gardner and N. H. MacKenzie (London, 1970), 70, 'Hurrahing in Harvest': 'And the azurous hung hills are his world-wielding shoulder'; Hopkins, *Sermons*..., ed. Devlin, 198: 'Satan, who is the κοσμοκράτωρ, the worldwielder, gave nature all an impulse of motion which should destroy human life'; *The Correspondence of Gerard Manley Hopkins and Richard Watson Dixon*, ed. C. C. Abbott (London, 1935), 53: 'In Nature is something that makes, builds up, and breeds... and over against this, also in Nature, something that unmakes or pulls to pieces...' How does GMH come, within the space of four years, to apply what is essentially the identical term to both the Saviour and Satan without detecting, so far as I can see, his own 'paradox and problem'?

The man himself... he was./Austin, *PP*, 78.

'working the dictionary'/ibid., 135. Not exactly as Austin prescribes ('quite a concise one will do', p. 134)!

'overriding'/*OED*: Hallam, *Constitutional History* (1876), I vi 349. 'The unconstitutional and usurped authority of the star-chamber over-rode every personal right.' Calhoun, *Works* (1874), III, 589. 'The Constitution must override the deeds of cession, whenever they come in conflict.'

148 '*insouciant* latitude'/Austin, *Sense and Sensibilia*, 58.

'could not bear... fantasy'/in Berlin and others, *Essays*..., 43, 27; Berlin, op. cit., 101.

'decent and comely order'/*Princeton Encyclopedia of Poetry and Poetics*, ed. A. Preminger, F. J. Warnke and O. B. Hardison Jnr, enlarged edition (London, 1975), 870.

'*parasitic*... joking'/Austin, *Words*, 22, 9.

'spiritual assumption... *bond*'/ibid., 10.

'flurried' or 'anxious to get off'/Austin, *PP*, 132.

'jargon... ordinary language'/ibid., 137.

'superstition... language'/ibid., 133.

'clear up... snares us'/ibid., 125, 137, 150.

'cheerfully subscribe... ideals'/ibid., 151 n. 1.

'the old Berkeleian... manifold"'/Austin, *Sense and Sensibilia*, 61.

'a_2... clumsily"]'/Austin, *PP*, 147 and n. 1.

149 Iris Murdoch/In *The Sovereignty of Good* (London, 1970), 25.
R. L. Nettleship/*Philosophical Remains of Richard Lewis Nettleship*, ed. A. C. Bradley, second edition (London, 1901), 138 ['There is much conjectural restoration in
this paragraph.']
'drama of reason'/*Collected Letters of Samuel Taylor Coleridge*, ed. Earl Leslie
Griggs, 6 vols (Oxford, 1956–71), III, 282.
Words . . . use it./Nettleship, *Philosophical Remains*, 136–7.
'LIVING POWERS'/Coleridge, *Aids to Reflection*, ed. cit., xvii.
'within the process'/Green, *Works*, III, 72.
'outside the process . . . developed'/ibid.
'ecstasy'/ibid., 79.
'transparency' of description/G. J. Warnock, *English Philosophy Since 1900*, second
edition (London, 1969), 102: '[Austin's] aim is not merely to describe the workings
of language . . . transparent.'
Kantian 'manifold' . . . tincture/Green, *Works*, III, 153 ('manifold of sense'), 79
('the world's manifold').
150 'malicious pleasure'/Berlin, *Personal Impressions*, 112.
'if the poet says . . . order'/Austin, *PP*, 228.
Coleridge observed . . ./*Biographia Literaria*, II, 116.
'linguistic phenomenology'/Austin, *PP*, 130 ['only that is rather a mouthful'].
'ontological manoeuvre'/John Crowe Ransom, *The World's Body*, paperback edition (Baton Rouge, 1968), 347: 'The critic should regard the poem as nothing short
of a desperate ontological or metaphysical manoeuvre.'
R. P. Blackmur . . . distress/In *Language as Gesture: Essays in Poetry* (London, 1954),
358.
Kenneth Burke/In *A Grammar of Motives* (New York, 1945), 491.
'satisfies the desire . . . end attained'/A. E. Teale, *Kantian Ethics* (London, 1951),
209.
'verbal precision' . . . watch-words'/*Biographia Literaria*, II, 116–17.
'*Prae-tendere* . . . the crime'./Austin, *PP*, 208.
'Nec te Lar . . . amicae . . .'/*Ovid in Six Volumes*, II, *The Art of Love and Other Poems*
with an English translation by J. H. Mozley (London, 1929, reprinted 1969), 194.
If 'crime' is too 'portentous', Mozley's 'weakness' is arguably not emphatic enough.
'that rather woolly word "imply"'/Austin, *PP*, 224.
'issued in ordinary circumstances'/Austin, *Words*, 22.
151 'the density . . . density'/John Crowe Ransom, *The New Criticism* (Norfolk, Con-
necticut, 1941), 79.
Those invocations/Hopkins, *Poems*, ed. cit., 58, 62, 57.
'Dim sadness . . . name'/*Wordsworth: Poetry and Prose*, selected by W. M. Merchant, third impression (London, 1969), 523 ('Resolution and Independence').
'A pleasurable . . . love'./ibid., 195 ('Michael').
'dogged in den'/Hopkins, *Poems*, ed. cit., 54 ('The Wreck of the Deutschland').
'dear and dogged man'./ibid., 91 ('Ribblesdale').
'heavy bodies'/*The Note-Books and Papers of Gerard Manley Hopkins*, ed. Humphry
House (London, 1937), 223: 'We may think of words as heavy bodies . . . Now
every visible palpable body has a centre of gravity . . . The centre of gravity is like
the accent of stress, the highspot like the accent of pitch . . .'
'most recondite and difficult'/Hopkins, *Sermons* . . . , ed. Devlin, 200.

'ordinary . . . beliefs'/Austin, *Words*, 22; Green, *Works*, III, 117, lxxiv; Berlin, op. cit., 101.

'the audible . . . company'/Fowler (ed.), *Locke's Conduct of the Understanding*, 98.

'*bona fide* perplexity'/Green, *Prolegomena to Ethics*, ed. A. C. Bradley, fifth edition (Oxford, 1907), 403.

'paradoxes and problems'/ Donne, *Iuuenilia: or certaine paradoxes and problemes*. 4° London, Printed by E. P[urslowe]. for Henry Seyle . . . 1633.

152 'impudently'/J. B. Leishman, *The Monarch of Wit* (London, 1951), 56: 'Such are the impudent paradoxes . . . of Ovid's *Amores* . . . Donne seems to have been the first to perceive what novel, surprising and shocking effects might be produced by exploiting the . . . *Amores*.'

'I did best . . .'/'You know my uttermost when it was best, and even then I did best when I had least truth for my subjects.' Quoted in John Donne, *The Divine Poems*, edited . . . by Helen Gardner, reprinted lithographically . . . from corrected sheets of the First Edition (Oxford, 1966), xxxvii.

'to lay him . . . wantonnesse'/Donne, *Devotions upon Emergent Occasions*, ed. John Sparrow (Cambridge, 1923), 12.

'Or . . . that?'/ibid., 68–9.

He argues . . . destruction'./ibid.

Marlowe's Faustus . . ./R. C. Bald, *John Donne: A Life* (Oxford, 1970), 73: 'Marlowe seems to have made the deepest impression on [Donne] . . . and it has been shown that recollections of the damnation of Faustus returned to haunt his imagination in after years.'

'thinking experience'/A. J. M. Milne, *The Social Philosophy of English Idealism* (London, 1962), 89.

'*Noble* parts . . . *misery*'./Donne, *Devotions*, ed. Sparrow, 70.

Nygren/Anders Nygren, *Agape and Eros*, Part II, 'The History of the Christian Idea of Love', vol. II, authorized translation by P. S. Watson (London, 1939), 267–8 n. 3.

153 'hazardous course' . . . purposes'/*OED*: 'compromise', *v* 8; 4; *sb* 5 *fig.*

the dyer's hand . . . infected hand/Shakespeare, Sonnet *110*; *OED*: 'Infect', *v* [ad. L *infect-*, ppl. stem of *inficere* to dip in, stain] . . . [cl. L. in neut. pl. *infectiva* dyes].

'curva voluntas'/ St Augustine, *Enarrationes in Psalmos*, 44, 17.

Peirce/*Collected Papers of Charles Sanders Peirce*, ed. C. Hartshorne, P. Weiss et al, 8 vols (Cambridge, Mass, 1931–58), I, 7. I owe this quotation to Randall Jarrell, *Poetry and the Age* (London, 1955), 194.

'we must consider . . . speech-act'/Austin, *Words*, 52.

'saying . . . customers'/Warnock, in Berlin and others, *Essays* . . . , 83.

As Warnock notes . . ./ibid. See Austin, *Words*, 57; *PP*, 229–30.

154 'When I asked . . . no words came out'/Julien Cornell, *The Trial of Ezra Pound: A Documented Account of the Treason Case by the Defendant's Lawyer* (London, 1967), 32.

'self-stultifying procedures'/Austin, *Words*, 51: 'Just as the purpose of assertion is defeated by an internal contradiction (in which we assimilate and contrast *at once* and so stultify the whole procedure), the purpose of a contract is defeated if we say "I promise and I ought not".' Cf. Cornell, op. cit., 41: 'While the doctors are agreed that he is to this extent mentally abnormal . . . I think it may fairly be said that any man of his genius would be regarded by a psychiatrist as abnormal.'

('A. . . . of course')/Cornell, op. cit., 166.

'All four experts . . . state.'/ibid., 45.

'the key . . . Confucius'/ibid., 157.

'peculiar . . . law'/ ibid., 46.

'was imprisoned . . . fact'/ibid., vii.

'there is . . . opposition'/Donald Hall, *Remembering Poets: Reminiscences and Opinions* (New York, 1978), 242.

'that stupid and suburban prejudice'/ibid., 185: Pound in conversation with Allen Ginsberg.

'insufficient desperation'/Richard Reid, 'Ezra Pound Asking', in *Agenda*, 17, nos. 3–4 – 18, no. 1 (3 issues) (Autumn-Winter-Spring 1979/80), pp. 171–86 (p. 172): 'Yet to charge Pound with banality would be to understate the disturbing quality of this text, its insufficient desperation.'

155 'solid entities'/Empson, *Complex Words*, 39: 'A word may become a sort of solid entity, able to direct opinion . . .'

'compacted doctrines'/ibid.

T. H. Green argued . . . itself'/Green, *Works*, III, 116.

'act of attention'; 'disinterested concentration of purpose'/Nettleship, *Philosophical Remains*, 135, 392.

'I guess I was off base all along'/Hall, op. cit., 148.

In a note . . . fatiguing/Cornell, op. cit., 73.

sounds magisterially Shelleyan/As Pound himself remarked, *Letters*, 249: '(This arrogance is not mine but Shelley's, and it is absolutely true . . .)'

logopoeia/*Literary Essays of Ezra Pound*, ed. T. S. Eliot (London, 1954), 25: 'LOGOPOEIA, "the dance of the intellect among words".'

Christopher Ricks's redefinition/In conversation with me.

'The poet's job . . . justice'/Pound, *Letters*, 366.

'the cultivation . . . law'/*Biographia Literaria*, II, 117.

156 'Fact is richer than diction'/Austin, *PP*, 143.

'I certainly omitted . . . open to me'/Pound, *Letters*, 310–11.

And now . . . Fama volet'./J. P. Sullivan, *Ezra Pound and Sextus Propertius: A Study in Creative Translation* (London, 1965), 170–1.

('Near . . . book-stall')/Pound, *Personae*, 220.

'imbecility'/Pound, *Letters*, 310: 'faced with the infinite and ineffable imbecility of the British Empire . . . faced with the infinite and ineffable imbecility of the Roman Empire.'

'exemplary instances . . . vocabulary'/Nathan A. Scott, Jnr, *The Poetry of Civic Virtue* (Philadelphia, 1976), x, 123, 135.

'redeeming . . . absurdity'/ibid, 102, 97.

'absolute closures of possibility'/ibid, 66.

157 'foundational' . . . 'primal reality'/ibid., 4.

'wanting the precept . . . what is'/Sidney, *Prose Works*, III, 14.

'dark and disputed . . . field'/Hopkins, *Sermons* . . . , ed. Devlin, 149–51.

'*the doing* be . . . so-and-so'/ibid.

'hantle of howlers'/Edwin Morgan, *Essays* (Cheadle Hulme, 1974), 255–76.

'Caesar . . . misagony'/ibid., 259–63.

'are hardly . . . imagination'/ibid., 259.

'It may seem . . . mind'/ibid., 260.

'the self unqualified in volition'/Silber, in Grene (ed.), op. cit., 206.
'murky . . . actuality'/Morgan, op. cit., 263, 258.
158 'first-grade . . . pun'/ibid., 257.
'ideas of guilt . . . interesting'/Murdoch, op. cit., 68.
'Verbalism . . . skill'/Pound, *Literary Essays*, 283.
'a sloppy and slobbering world'/Pound, *Letters*, 378.
'the transition . . . Pound/C. H. Sisson, *English Poetry 1900–1950: An Assessment*
(London, 1971), 102.
'I never . . . God damn it'/Cornell, op. cit., 192.
'the tyro . . . dangerous'/Pound, *Literary Essays*, 283.
'the finest wrought'/ibid.
'the crime . . . writing'/Cornell, op. cit., 195.
'*by his being . . . his signature*'/Austin, *Words*, 60.
159 'a verdictive . . . exercitives'/ibid., 152.
'Now comes . . . incompetent person . . .'/Cornell, op. cit., 61.
'And when . . . finished with the subject.'/Pound, *Letters*, 347.

Index